Component Models and Systems
for Grid Applications

T0205333

Component Models and Systems for Grid Applications

Proceedings of the Workshop on Component Models and Systems for Grid Applications held June 26, 2004 in Saint Malo, France.

Edited by

Vladimir Getov
University of Westminster
London, United Kingdom

Thilo Kielmann
Vrije Universiteit
Amsterdam, The Netherlands

 Springer

Dr. Vladimir Getov
University of Westminster
London
UNITED KINGDOM

Thilo Kielmann
Vrije Universiteit
Amsterdam
NETHERLANDS

Library of Congress Cataloging-in-Publication Data

A C.I.P. Catalogue record for this book is available from the Library of Congress.

Component Models and Systems for Grid Applications / edited by
Vladimir Getov, Thilo Kielmann
    p.cm.
        Includes bibliographical references and index.

ISBN 978-1-4419-3614-1
e-ISBN 978-0-387-23352-9

Printed on acid-free paper.

©2010 Springer Science+Business Media, Inc.
All rights reserved. This work may not be translated or copied in whole or in part without
the written permission of the publisher (Springer Science+Business Media, Inc., 233 Spring
Street, New York, NY 10013, USA), except for brief excerpts in connection with reviews or
scholarly analysis. Use in connection with any form of information storage and retrieval,
electronic adaptation, computer software, or by similar or dissimilar methodology now
know or hereafter developed is forbidden.
The use in this publication of trade names, trademarks, service marks and similar terms,
even if the are not identified as such, is not to be taken as an expression of opinion as to
whether or not they are subject to proprietary rights.

Printed in the United States of America.

9 8 7 6 5 4 3 2 1

springeronline.com

Contents

Foreword

Grid computing emerged in the mid-nineties as a new high-performance computing paradigm for scientific and engineering applications. At this early stage, Grids focused mainly on metacomputing – sharing computing resources over the Internet. The current architectural extensions to the Grid concept are still not sufficient to meet the broader requirements, such as those derived from business or heavy-duty scientific needs. These requirements include scalability, robustness, dynamic reconfigurability, self-healing, high integrity, business-strength security and trust, low-effort-threshold end-user interfaces, homogeneous and transparent access to both heterogeneous data and processing resources.

Therefore, research and development in the area of Grid technologies is a key strategic objective, not only for the scientific community, but also for society as a whole. Hence, it has been given high priority and attention by a number of funding bodies and agencies world-wide. The European "Information Society Technologies" (IST) research and development programme[1] in particular has been giving high priority to Grid-related research for several years now with a large number of recent and currently active projects. Amongst them, CoreGRID aims at strengthening and advancing long-term research, knowledge transfer and integration in the area of Grid and Peer-to-Peer technologies. It is a network of excellence – a new instrument within the 6th Framework Programme – to ensure progressive evolution and durable integration of the European Grid research community. In order to achieve this objective, CoreGRID brings together a critical mass of well-established researchers and doctoral students from forty-two institutions who have developed an ambitious joint programme of activities. This programme is structured around six complementary research areas that have been selected on the basis of their strategic importance, research challenges, and recognised European expertise to develop next generation Grid middleware, namely: knowledge and data management; programming models; system architecture; information and monitoring services; resource management and scheduling; Grid systems, tools and problem solving environments.

---

[1] http://www.cordis.lu/ist/home.html

Besides other activities, the CoreGRID working plan envisages the organisation of a number of international peer-reviewed workshops and conferences in this exciting and rapidly developing area. The present proceedings are the first volume of a series of publications that will be edited by CoreGRID researchers. They reflect the preparation of the CoreGRID proposal, which evolved out of several working meetings. These meetings fostered relationships between the forty-two partners and were the occasion to identify a number of important research issues related to future Grid systems. Vladimir Getov and Thilo Kielmann, the editors of this volume, took a pro-active role in this process and also co-chaired the Workshop on Component Models and Systems for Grid Applications, as part of the ACM ICS 2004 conference. This is a research field of growing importance addressing the problem of simplifying the inherent complexity of both Grid middleware and Grid applications. The workshop outlined some of the most interesting research topics covered during the preparation phase of the project. It is, therefore, not surprising that the first volume of the CoreGRID series of proceedings is a good selection of high-quality articles, which marks not only the start of the series but also the start of the CoreGRID project itself. On behalf of all CoreGRID partners, I would like to express my gratitude to the editors and to all contributors.

Thierry Priol, CoreGRID Scientific Co-ordinator

Preface

Grids are modern computer systems that enable access to and the sharing of geographically distributed heterogeneous resources, such as computers, knowledge, sensors, and instruments for solving large-scale and complex problems in science and commerce. Through virtualisation of these resources, Grids will become much easier to program and to use. The present capabilities of Grids are being largely demonstrated through a variety of representative e-science applications, but the ultimate commercial benefits will come from their potential to revolutionise tomorrow's Internet. Further research results and the promotion of early adoption by industry are crucial to move Grids from their current, pioneering stage in science to widespread use in both business and industry.

It is believed that the long-term evolution of Grids depends on the development and the significant advances in three different but complementary areas which are vital for building next generation Grids[1]:

- Architecture – this area envisions the Grid as a structural entity with a collection of capabilities and properties, which are critical in providing an indication of the scale expected from future Grids in terms of numbers, geography, and administrative domains.

- Software – this area focuses on how Grid programming will look, which algorithmic problems have to be solved and which constraints have to be observed in order to build an infrastructure as described by the Grid architecture.

- Applications – this area develops a vision of how the Grid could be deployed in the everyday life of individuals, businesses, and organisations.

Reflecting this new reality, the Workshop on Component Models and Systems for Grid Applications[2], organized by members of the EU CoreGRID project[3], was established to bring together researchers working in this field.

---

[1] ftp://ftp.cordis.lu/pub/ist/docs/ngg_eg_final.pdf
[2] http://www.cs.vu.nl/CMSGA/
[3] http://www.coregrid.org

The programming of Grid applications is highly complex due to the scale, heterogeneity, and dynamicity of Grid infrastructures. Whereas Grid middleware has made considerable progress recently, Grid application programs are still being developed mostly without support from advanced programming models and environments. Therefore, the development of component models and their integration into component architectures and systems has been recognised as a topic of high priority in order to fill this gap. The goal of the workshop was to bring together researchers in this exciting field of growing importance. Indeed, attendees with expertise in a variety of topics arrived in Saint Malo at the end of June 2004 to discuss the state-of-the-art and the future of this field. The topics covered by the workshop were grouped into four sessions starting with a key-note talk by Dennis Gannon.

The presentations during the workshop created an inspiring environment for the final panel discussion, entitled *Objects, Components, Services, and Peers: Can we handle all of them in one global system?* and moderated by Domenico Laforenza from the Italian National Research Council. After his thought-provoking introduction, the panelists (Dennis Gannon, Thierry Priol, Vladimir Getov, and Marco Danelutto) contributed their points of view. A lively discussion spun off as a result of their short presentations addressing the following main open questions and issues:

- How can we integrate the currently independent communities that are dealing with Grid computing, peer-to-peer systems, networks, and coordination models?

- WSRF is recognized as rather low level, WSDL is human-readable, but not understandable. What is the right level of abstraction for Grid components and services?

- What will be the "killer application" of Grids? Will it be the seamless computing, as envisaged by the power grid analogy?

- How can the Grid be made as simple to use as the Web? Will the simple user interface finally attract the "killer application"?

Of course, a panel discussion can only identify such questions; giving the right answers will be the task of ongoing research. However, the work presented at the workshop and collected in this book provides an excellent starting point to tackle these issues.

We have organized this book in three parts. Part I is devoted to *Application-Oriented Designs* including contributions on the development methodology for building component-based Grid applications. Part II explores the *Middleware Architecture* area with chapters emphasizing hierarchical infrastructures, workflow modelling environments, interoperability, and portable Grid engines.

Finally, Part III deals with *Communication Frameworks* addressing dynamic self-adaptation, higher-order components, and collective operations.

Finally, we would like to thank all the participants for their contributions to making the workshop a resounding success; the organisers of the 18th Annual ACM International Conference on Supercomputing for their professional support in the organization; the workshop program committee for reviewing the submissions; and, last but not least, all the authors that contributed articles for publication in this volume.

Our thanks also go to the European Commission for sponsoring under grant number 004265 this first volume of the CoreGRID project series of publications.

Vladimir Getov and Thilo Kielmann

Contributing Authors

Ali Afzal London e-Science Centre, Imperial College London, South Kensington Campus, London SW7 2AZ, UK (lesc-staff@doc.ic.ac.uk)

Marco Aldinucci Istituto di Scienza e Tecnologie dell'Informazione, ISTI-CNR, Via Buonarroti, 2 - 56127 Pisa, Italy (Marco.Aldinucci@isti.cnr.it)

Martin Alt Institut für Informatik, University of Münster, Einsteinstraße 62, 48149 Münster, Germany (mnalt@math.uni-muenster.de)

Françoise André IRISA/Université de Rennes 1, Campus universitaire de Beaulieu, 35042 Rennes cedex, France (Francoise.Andre@irisa.fr)

Bartosz Baliś Institute of Computer Science, AGH, al. Mickiewicza 30, 30-059 Kraków, Poland (balis@uci.agh.edu.pl)

Purushotham Bangalore Department of Computer and Information Sciences, University of Alabama at Birmingham, Birmingham, AL 35294, USA (puri@cis.uab.edu)

Françoise Baude INRIA Sophia Antipolis, CNRS - I3S - University of Nice Sophia-Antipolis, 2004, Route des Lucioles, BP 93, F-06902 Sophia-Antipolis Cedex - France (Francoise.Baude@sophia.inria.fr)

Marian Bubak Institute of Computer Science, AGH, al. Mickiewicza 30, 30-059 Kraków, Poland (bubak@uci.agh.edu.pl)

Jérémy Buisson IRISA/INSA, Campus universitaire de Beaulieu, 35042 Rennes cedex, France (Jeremy.Buisson@irisa.fr)

Sonia Campa Dept. of Computer Science, University of Pisa, Via Buonarroti, 2 - 56127 Pisa, Italy (campa@di.unipi.it)

Denis Caromel INRIA Sophia Antipolis, CNRS - I3S - University of Nice Sophia-Antipolis, 2004, Route des Lucioles, BP 93, F-06902 Sophia-Antipolis Cedex - France (Denis.Caromel@sophia.inria.fr)

Jeremy Cohen London e-Science Centre, Imperial College London, South Kensington Campus, London SW7 2AZ, UK (lesc-staff@doc.ic.ac.uk)

Massimo Coppola Istituto di Scienza e Tecnologie dell'Informazione, ISTI-CNR, Via Buonarroti, 2 - 56127 Pisa, Italy (Massimo.Coppola@isti.cnr.it)

Marco Danelutto Dept. of Computer Science, University of Pisa, Via Buonar-roti, 2 - 56127 Pisa, Italy (marcod@di.unipi.it)

John Darlington London e-Science Centre, Imperial College London, South Kensington Campus, London SW7 2AZ, UK (lesc-director@doc.ic.ac.uk)

Jan Dünnweber Institut für Informatik, University of Münster, Einsteinstraße 62, 48149 Münster, Germany (duennweb@math.uni-muenster.de)

Liang Fang Department of Computer Science, Indiana University, Blooming-ton, IN 47401, USA (lifang@cs.indiana.edu)

Nathalie Furmento London e-Science Centre, Imperial College London, South Kensington Campus, London SW7 2AZ, UK (lesc-staff@doc.ic.ac.uk)

Dennis Gannon Department of Computer Science, Indiana University, Bloom-ington, IN 47401, USA (gannon@cs.indiana.edu)

Vladimir Getov Harrow School of Computer Science, University of Westmin-ster, Watford Rd, Northwick Park, Harrow, London HA1 3TP, UK (V.S.Getov@westminster.ac.uk)

Sergei Gorlatch Institut für Informatik, University of Münster, Einsteinstraße 62, 48149 Münster, Germany (gorlatch@math.uni-muenster.de)

Jeff Gray Department of Computer and Information Sciences, University of Alabama at Birmingham, Birmingham, AL 35294, USA (gray@cis.uab.edu)

Murtaza Gulamali London e-Science Centre, Imperial College London, South Kensington Campus, London SW7 2AZ, UK (lesc-staff@doc.ic.ac.uk)

Francisco Hernández Department of Computer and Information Sciences, University of Alabama at Birmingham, Birmingham, AL 35294, USA (hernandf@cis.uab.edu)

Stavros Isaiadis Harrow School of Computer Science, University of Westminster, Watford Rd, Northwick Park, Harrow, London HA1 3TP, UK (S.Isaiadis@westminster.ac.uk)

Gopi Kandaswamy Department of Computer Science, Indiana University, Bloomington, IN 47401, USA (gkandasw@cs.indiana.edu)

Sriram Krishnan Department of Computer Science, Indiana University, Bloomington, IN 47401, USA (srikish@cs.indiana.edu)

Domenico Laforenza Istituto di Scienza e Tecnologie dell'Informazione, ISTI-CNR, Via G. Moruzzi, 1 - 56126 Pisa, Italy (Domenico.Laforenza@isti.cnr.it)

Bruce Long Harrow School of Computer Science, University of Westminster, Watford Rd, Northwick Park, Harrow, London HA1 3TP, UK (B.D.Long@westminster.ac.uk)

Andrew Lumsdaine Open Systems Laboratory, Indiana University, 501 N. Morton St., Bloomington, IN 47404, USA (lums@open-mpi.org)

Anthony Mayer London e-Science Centre, Imperial College London, South Kensington Campus, London SW7 2AZ, UK (lesc-staff@doc.ic.ac.uk)

Steve McGough London e-Science Centre, Imperial College London, South Kensington Campus, London SW7 2AZ, UK (lesc-staff@doc.ic.ac.uk)

Matthieu Morel INRIA Sophia Antipolis, CNRS - I3S - University of Nice Sophia-Antipolis, 2004, Route des Lucioles, BP 93, F-06902 Sophia-Antipolis Cedex - France (Matthieu.Morel@sophia.inria.fr)

Jens Müller Institut für Informatik, University of Münster, Einsteinstraße 62, 48149 Münster, Germany (jmueller@math.uni-muenster.de)

Steven Newhouse London e-Science Centre, Imperial College London, South Kensington Campus, London SW7 2AZ, UK (lesc-staff@doc.ic.ac.uk)

Jean-Louis Pazat IRISA/INSA, Campus universitaire de Beaulieu, 35042 Rennes cedex, France (Jean-Louis.Pazat@irisa.fr)

Diego Puppin Istituto di Scienza e Tecnologie dell'Informazione, ISTI-CNR, Pisa, Italy and Dept. of Computer Science, University of Pisa, Via Buonarroti, 2 - 56127 Pisa, Italy (Diego.Puppin@isti.cnr.it)

Kevin Reilly Department of Computer and Information Sciences, University of Alabama at Birmingham, Birmingham, AL 35294, USA (reilly@cis.uab.edu)

Luca Scarponi Dept. of Computer Science, University of Pisa, Via Buonarroti, 2 - 56127 Pisa, Italy (scarponi@di.unipi.it)

Alexander Slominski Department of Computer Science, Indiana University, Bloomington, IN 47401, USA (aslom@cs.indiana.edu)

Jeffrey M. Squyres Open Systems Laboratory, Indiana University, 501 N. Morton St., Bloomington, IN 47404, USA (jsquyres@open-mpi.org)

Jeyarajan Thiyagalingam Harrow School of Computer Science, University of Westminster, Watford Rd, Northwick Park, Harrow, London HA1 3TP, UK (T.Jeyan@westminster.ac.uk)

Marco Vanneschi Dept. of Computer Science, University of Pisa, Via Buonarroti, 2 - 56127 Pisa, Italy (vannesch@di.unipi.it)

Michal Wegiel   Institute of Computer Science, AGH, al. Mickiewicza 30, 30-059 Kraków, Poland (mwegiel@uci.agh.edu.pl)

Laurie Young London e-Science Centre, Imperial College London, South Kensington Campus, London SW7 2AZ, UK (lesc-staff@doc.ic.ac.uk)

Corrado Zoccolo Dept. of Computer Science, University of Pisa, Via Buonarroti, 2 - 56127 Pisa, Italy (zoccolo@di.unipi.it)

I

APPLICATION-ORIENTED DESIGNS

BUILDING APPLICATIONS FROM A WEB SERVICE BASED COMPONENT ARCHITECTURE

Dennis Gannon, Sriram Krishnan, Alexander Slominski, Gopi Kandaswamy, and Liang Fang
Department of Computer Science
Indiana University
Bloomington, IN, USA
gannon@cs.indiana.edu
srikish@cs.indiana.edu
aslom@cs.indiana.edu
gkandasw@cs.indiana.edu
lifang@cs.indiana.edu

Abstract   This chapter describes an approach to building large-scale, distributed applications based on a software component composition model that allows web services to be used as the basic units. The approach extends the Common Component Architecture used in many parallel supercomputer applications, from static composition of directly coupled processes to a system that incorporates mediated workflow between remote services. The system also allows legacy applications to be easily wrapped as a component and executed from a service factory. We motivate the work in terms of a large, distributed application for modeling severe storms. The entire system is based on a three-level architecture with a portal providing the user interface, a set of security and factory service utilities in the middle and the application services and components in the back-end.

Keywords: Web services, software component architectures, legacy applications, portals, XCAT3

1.   Introduction

The goal of building software from pluggable components is not new. Unix pipes allowed simple linear composition of programs based on composing an output stream with an input stream. In the area of computer visualization systems like AVS [9] allowed users to put together complex graphics applications by composing graphs of simple filter and rendering components. The Common Component Architecture (CCA) [4] is a modern component system that is used to build large supercomputer simulations by coupling together components such as linear solvers and boundary condition evaluators. Many other component systems exist they differ primarily in the way individual components are required to behave and in the semantics of the method of composition. Most use a variation of the "inversion of control" pattern which is based on extracting the control of the overall application and placing it in surrounding framework.

Grid systems are distributed frameworks for sharing resources among the membership of a virtual organization. Currently Grids are designed as a collection of common services which provide security, data management, application execution scheduling, notification and logging, policy expression and system monitoring. These services are often implemented as web services and many Grid applications can be constructed by composing collections of basic services and "atomic" application components.

For example, the LEAD project [10] is trying to build a distributed cyber infrastructure powerful enough to enable the "better than real-time" prediction of mesoscale weather events such as tornadoes. One of the goals of LEAD is to allow a scientist to compose applications that are advanced Grid measurement and prediction scenarios of various types. For example, an application may allow a user to select a region of land such as a state or a few counties, and a data query, such as current radar data. From this information, the application begins a data mining monitoring process which searches the radar data in that region for anomalous behavior. When strange conditions are detected, compute resources are allocated and a series of simulations are launched. The simulations are monitored to see how they align with the real weather and those that are not consistently tracking reality are terminated. Addition resources may be applied to those that are accurate. For example, the radar array be asked to provide more detail so the resolution of the simulation can be increased. This type of application is a good example of a dynamic workflow that requires an extensive distributed framework of services to be composed. Specifically, the services include

- Metadata catalog services for extracting information about past storm histories and available instrument data streams for a given geographic region.

- A data mining service that can be configured to mine the instrument data.

- Resource allocators for both space and computation

- Simulation instance factories that can be used to launch version of a simulation with different parameters.

- Visualization services that can convert the output of simulations into movies and other display data the user wants.

- Logging services so the user can keep track of what is going on.

In the sections that follow we will outline a service composition model that allows these service components to be composed by scientists to enact specific experimental prediction scenarios. The system is based on a three level architecture. The user interacts with the system by means of a Grid portal that forms the first level. In the second level we have the security framework and application factories. At the third level are the specific service instances that participate in computation.

2.   The Portal

The Grid portal is based on the Open Grid Computing Environment (OGCE) [23] framework. But it could also use any JSR-168 compliant portal such as the GridSphere [22]. The OGCE portal allows the user to interact Grid applications and services from a standard web browser. The portal provides each user with a context of resources including a proxy identity certificate to allow the portal to authenticate with remote grid services on behalf of the user. The portal provides tools for defining geographic regions, querying and searching metadata catalogs, checking job execution logs, cataloging experimental results and defining workflows with simple graphical tools.

The portal represents the top layer of the grid stack. The bottom layer of the stack is a set of shared resources. These may be real (computers, databases, instruments) or virtual (documents, name spaces, ontoloties). Above the resource layer we have the Open Grid Services Infrastructure (OGSI) [13] or the Web Service Resource Framework (WSRF) [14], which is a set of basic web service abstractions designed to provide a standard mechanism for describing resources.

From our perspective, a set of services that provide a mechanism for communicating "events", such as WS-Notification, is critical. We will return to this issue below.

Built upon these foundations we have the Open Grid Service Architecture which is the federation of services that define the core grid platform. Finally, as shown in Figure 1, the portal build upon these services to create the application level services it needs.

While the portal provides several other workflow composition tools, such as an interface to upload Dagman Condor scripts, the graphical component

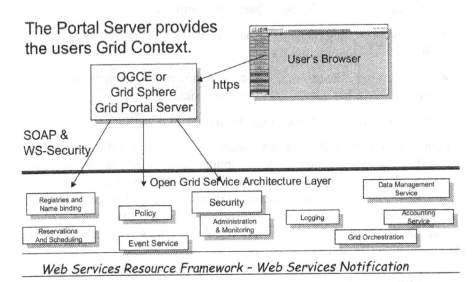

Figure 1.   The layered organization of a Grid. The portal is the user's access point.

composition tool is the primary mechanism for building applications based on web services and CCA components. The user interface 2 is a simple "drop and drag" composer that is similar to the standard CCA composer or the Scirun II composer[1]. It is also similar to the Triana [7] and Kepler [18] interfaces.

The primary difference between our composer and others lies in the back end. The OGCE composer allows both web service and CCA components to be integrated into a single application. While this work is still in progress and not yet released, it is based on a compiler that translates the graphical specification of the application into a standard workflow language. We will return to this topic later in this chapter.

There is one more important point to make about the composer: it is an example of how a Grid services provides an interface to the user through the portal. Our model for doing this is similar to the WSRP specification [20], but it takes advantage of the structure of a Grid service. For Grid services that have user interfaces, we have defined a "standard" service data element (in OGSI terms) or resource (in WSRF) terms that provides the URL to load the GUI for that service. This GUI may be an applet (as in the case of the composer) or it may be an xhtml document with imbedded java script.

---

[1]Scirun [24], from the University of Utah, is also based on CCA.

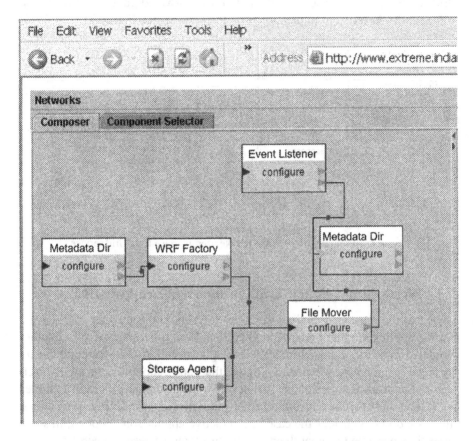

Figure 2. Component Composer Interface. This is the standard way most component systems use to assemble small workflows. This version is an applet that will be distributed as part of OGCE.

There are two difficult problems that must be solved when you want a user to interact with a service through a remote client interface. What do you do when the Grid service is behind a firewall? What do you when the Grid service requires the user to use WS-Security when it talks to the service? The user may have the interface, but not their own X.509 identity certificate. The solution to both of these problems is to filter all transaction between the service and the user through the portal, which we assume is a gateway through the firewall and also has access to the user's proxy certificiate. The protocol is as follows. When the user discovers the Grid service in the portal directory, a special servlet in the portal fetches the interface and passes it to the user. The user interact with the GUI which sends user commands back though the browser HTTPS connection to the portal. The portal then forwards those commands back to the

Grid service using the user's credentials for the XML digital signature in the WS-Security-based communication with the service (see Figure 3).

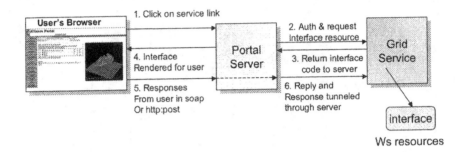

Figure 3.   Protocol for loading a remote user interface for a Grid service.

3.   Wrapping Legacy Applications as Components

One of the most frequently asked questions about this system is, "I have a Fortran application that I cannot modify. Can I use it in this system?" In other words, in what way can we turn a legacy application into a service component? By legacy application, we mean a program that can be submitted through a batch queue. For example, a script that that inserts some text into a program, compiles it and then runs it while consuming some input files and producing some output files. We do not consider the case of interactive applications, though, in some cases, it is possible to threat them in the same way we describe here.

Our approach is to use the factory pattern. We will build a service which is capable of starting an instance of the application on behalf of the service client. The input to the factory is either through a direct web service call, or through a user interface as described in the previous section. The input consists of any configuration parameters that the factory needs to start the job running.

There are two problems that must be solved. First, it is not hard for a web-service programming professional to write a web service to build an application factory. But it is often the case that the person who wishes to build such a service is the scientific programmer who is responsible for the deployment and testing of the application. This person is not a web-services programming expert. So can we automatically generate the factory for an application from a specification provided by the application provider?

The second problem involves security. One problem with sharing legacy applications is deployment. An application may run in one user's environment, but it may take a substantial effort to deploy it in another user's environment. Furthermore once an application leaves the developers control, that developer now has a version control problem. This is one motivation for provide the application as a service rather than as a program. But if an application provider creates an application factory that can instantiate a running instance of the application for somebody else, then how do we provide the authorization control mechanism that determines exactly who is allowed to run the application?

Taking the first question, we note that we have built an *Factory Service Generator* into the portal server. This allows the application provider to automatically generate a factory service from a relatively simple xml specification. The user need only provide:

- a script that can execute the application. Any needed input files should appear as filename parameters to the script. In addition we assume that the script take a "jobname" parameter so that it can create a private working directory for each run of the program. (The factory service assumes it can run multiple instances of the application concurrently so care must be taken that intermediate and final files are not overwritten.)

- an XML file that describes the application and the input parameters and other annotations to be placed on the user interface to the factory.

Given these two items, the portal can automatically generate a factory and start it running on behalf of the application providing user. When invoked the factory simply executes the script. There are no restrictions on this script other than the those described above. In many cases, we have the script run another web service which is a transient instance of a service that is dedicated to one user. In other cases, we have the script execute a complete workflow. In any case it is very useful if the script has the capability of sending event notifications such as "job complete", "output file is at URL ../jobname/filename" or "job failed because...". Python, GridAnt [26] and other high level scripting tools have this capability. We will describe how this is used later.

Once the factory is running, the provider must decide who is allowed to invoke it. The solution we use is based on capabilities. Each application provider supplies a list of individuals or groups that he or she will allow to run his or her application factory. The portal capability manager will then, for each user and group, create a signed XML capability document that says this individual or group has permission to execute the application factories "create instance" methods. When one of these users logs into the portal, the portal server loads the proxy cert for the user, which is then used to load that user's capabilities (see Figure 4). If the user invokes the factory service, the appropriate capability is added to the SOAP header for the service request. The factory service verifies that the request is authentic and that the capability is authentic and that the requestor is the same person as the capability certificate.

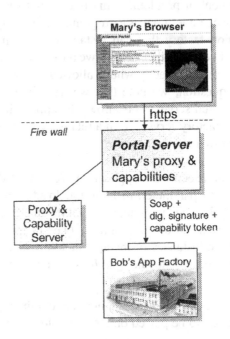

Figure 4.   Capability-based authorization protocol for access to factory services.

4.   XCAT3 - Integrating CCA Components and Grid Services

We briefly introduce CCA [4] and OGSI before presenting the architecture of the XCAT3 [19] framework in the following subsections.

4.1   Common Component Architecture

CCA is defined by a set of framework services, and the definitions of the components that use them. Each component communicates with other components by a system of *ports*. Ports are defined by a type system that is expressed in the Scientific Interface Definition Language (SIDL) [11]. SIDL is similar in nature to WSDL, but differs in the support it provides to data structures common to scientific computing. There are two types of CCA ports:

- **Provides Ports** are the services offered by the component. Each provides port implements an interfaced defined in SIDL.

- **Uses Ports** are the hooks that enable a component to use a service provided by another component. Uses ports are stubs that a component *uses* to invoke the services *provided* by another component. Uses ports are also defined in SIDL.

A uses port of one component can be connected to a provides port of another component as long as they implement the same SIDL interface. Connections between uses and provides ports are made at runtime. A component needs to execute a `getPort` statement to grab the most recent reference to provider, and a `releasePort` when it has finished using it. The get/release semantics of component connections enable the framework to infer if any port calls are being made at any point in time, and also enable the connections to be changed dynamically.

Apart from uses and provides ports, a component also implements a *ComponentID* interface that has methods that uniquely indentify the component, and provide metadata about it. CCA also defines a *Builder* service for creation and composition of these components.

4.2   Grid Services

The Open Grid Services Infrastructure extends the Web services model by defining a special set of service properties and behaviors for stateful Grid services. Some of the key features of OGSI that separate Grid services from simple Web services are:

- **Multiple level naming:** OGSI separates a logical service name from a service reference. A Grid Service Handle (GSH) provides an immutable name for a service, while a Grid Service Reference (GSR) provides a precise description of how to reach a service instance on a network, e.g., a WSDL reference. A GSH can be bound to different GSRs over time.

- **Dynamic Service Introspection:** Grid services can expose metadata to the outside world through the use of Service Data Elements (SDE), which are XML fragments that are described by a Service Data Descriptor (SDD). SDEs can be queried by name or type, and can be used to notify state changes to clients.

- **Standardized ports:** Every Grid service implements a *GridService* port, which provides operations to query for SDEs, and manage lifetime of the Grid service. OGSI also specifies standard ports for creation, discovery, and handle resolution.

Recently, OGSI has been superceded by the WSRF [14] proposal, which also addresses the above issues for stateful Grid services, but tries to integrate them better with the current Web service standards.

4.3   XCAT3 Architecture

Currently, the XCAT3 framework is implemented in Java, and we plan to implement a C++ version that is interoperable with the former. In XCAT3, we

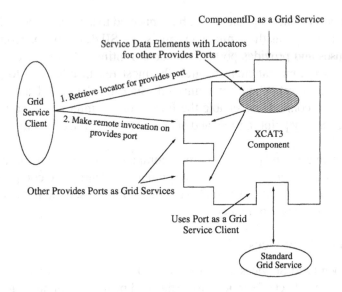

Figure 5.   Every XCAT3 component port is a Grid service. It contains SDEs with locators for all provides ports, which are also Grid services themselves.

implement the CCA specification in the context of Grid services. To that end, some of the key features of XCAT3 are:

- **Ports as Grid services:** As per the CCA specification, one component can have more than one provides port of the same type. Simple Grid and Web services allow multiple ports of the same portType; however, multiple bindings of the same port are semantically equivalent. Hence, the same operation on different ports of the same type affect the service in exactly the same way. However, unlike Web service ports, ports in CCA are designed to be stateful. Hence, every provides port in XCAT3 is implemented as a separate Grid service 5. The consequence of this is that every provides port inherits multiple level naming from the OGSI specification, and this enables the ports to be location independent. Additionally, any Grid service that is compliant with the OGSI specification can serve as a provides port.

- **ComponentID as a Grid service:** The ComponentID, as specified by the CCA specification, is also implemented as a Grid service. It exposes handles and references of all the provides ports that a component contains, and thus acts as a manager for the component. Users can query a component for the types of services provided via the ComponentID, and connect to them directly.

Some of the other useful services in the XCAT3 framework are:

- **Builder Service:** As mentioned before, the Builder service defines methods for component creation and composition. We allow remote instantiation of components via ssh or Globus GRAM provided by the Java CoG [3] kit. For composition purposes, the Builder service provides connect and disconnect methods for connecting and disconnecting a uses port to a provides port respectively. Once the ports have been connected, all communication between them is via Web service invocations provided by the XSOAP [25] toolkit.

- **Handle Resolver:** Since we employ multiple level naming for our ports and ComponentIDs, we need to use a handle resolution mechanism that translates a GSH to a GSR. This is provided by the Handle Resolver service. The Handle Resolver, as other Grid services in the XCAT3 framework, is implemented using the GSX [16] toolkit, which provides a lightweight implementation of the OGSI specification.

5.   An Example

As an example that integrates most of these ideas together, we consider an application that is part of a much larger LEAD scenario. Much of LEAD requires large simulations based on weather data input. The output of these simulations consists of data fields that represent severe thunder storms and tornadoes. It is useful to have a tool that can generate visualizations from these simulation outputs. We have an application factory that launches a OGRE script [2] for which sets up the LAM MPI on a cluster and runs a parallel rendering program on the output from the Weather Research and Forecasting (WRF) job. Once the parallel rendering is complete, it launches a conversion program which translates the rendered output to a GIF movie that can be viewed from the browser. OGRE scripts are capable of publishing events into the notification system (which is currently being converted to work with WS-Notification). An event listener, listens for all events published under the topic defined by the name of this OGRE execution. These events are logged into a special directory service visible from the portal. The entire workflow is depicted in Figure 6. (Note that the workflow for this example predates the composer tool described earlier. The workflow was hand crafted and not generated from the picture.)

The user can discover the status of each execution of this workflow by going to the portal "Grid Context" directory service. There entry for each execution looks like a directory which contains a list of all parameters used to launch the workflow, the log of execution events and a reference to the output GIF movie. As shown in Figure 7, selecting "status" displays the log of events received. In Figure 8, we see that selecting "results" will run the GIF movie in the right hand window.

WRF Simulation
Portal Directory
OGRE job Factory
File Mover
OGRE instance

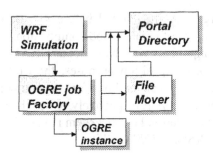

Figure 6. Complete distributed WRF output animation application. The output from the WRF simulation is used as input to the OGRE animation script. All components log events to the event channel. The event listener captures them and pushes them to the directory service.

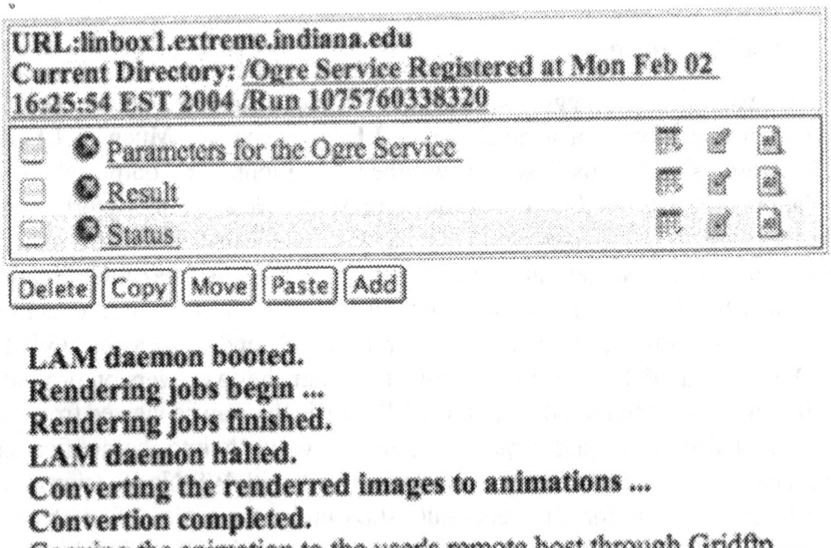

URL:linbox1.extreme.indiana.edu
Current Directory: /Ogre Service Registered at Mon Feb 02 16:25:54 EST 2004 /Run 1075760338320
Parameters for the Ogre Service
Result
Status
Delete
Copy
Move
Paste
Add
LAM daemon booted.
Rendering jobs begin ...
Rendering jobs finished.
LAM daemon halted.
Converting the renderred images to animations ...
Convertion completed.
Copying the animation to the user's remote host through Gridftp ...

Figure 7. A Directory service record for this execution of the workflow. Selecting "Status" shows the even log.

6. Conclusion

This chapter has illustrated a three level architecture for distributed Grid applications. At the top level we have a Grid portal which contains a suite of tools for creating grid services and composing application from them. This portal provides the secure interface to the back end Grid which may run behind a

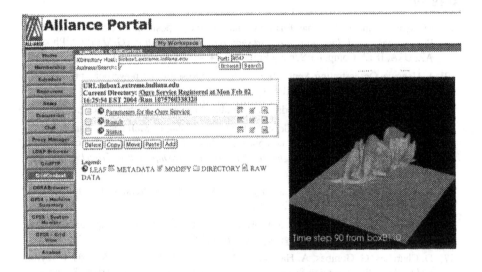

Figure 8. Selecting "Results" shows the output movie on the right.

firewall. The middle tier is a set of services that support our Grid applications. These include security services, such as a proxy certificate repository, an authorization system based on capability tokens, directory services and a notification system. One important tools is a factory service generator that allows users to describe a legacy application and the portal will generate a factory service to access and launch instances of that application. The back end is composed of the application resources and services.

We also describe how we transformed the Common Component Architecture into a Grid service-based framework. This allow CCA components to be used as Grid services and composed into service-based workflows. It also allows regular web services to be integrated into CCA distributed applications.

In the year ahead we will on building applications from services and CCA components. While much has been done, we feel there is much more to be learned and the most important discoveries will come from applying this technology to real problems.

Acknowledgments

We would like to thank our collaborators on the LEAD project for their help in understanding the issues, and the NSF and DOE for their support of the projects that funded this work.

References

[1] M. Agarwal, and M. Parashar. Enabling Autonomic Compositions in Grid Environments. Proceedings of *4th International Workshop on Grid Computing (Grid 2003)*, Phoenix, AZ, USA, IEEE Computer Society Press, pp. 34–41, November 2003.

[2] J. Alameda, Orchestrating Applications on Remote Resources, a powerpoint presentation, www.grids-center.org/train/GRIDS-Alameda.ppt.

[3] Argonne National Lab. Commodity Grid Toolkit. http://www.globus.org/cog. 2004.

[4] R. Armstrong, D. Gannon, A. Geist, K. Keahey, S. Kohn, L. McInnes, S. Parker, and B. Smolinski. Towards a common component architecture for high performance scientific computing. In Proceedings of *Eighth IEEE International Symposium on High Performance Distributed Computing*, 1998.

[5] Business Process Execution Language for Web Services Version 1.1. http://www-106.ibm.com/developerworks/library/ws-bpel/

[6] H. Casanova and J. Dongarra. NetSolve: a network server for solving computational science problems. In Proceedings *Supercomputing (SC 96)*.

[7] D. Churches, G. Gombas, A. Harrison, J. Maassen, C. Robinson, M. Shields, I. Taylor, and I. Wang. Programming Scientific and Distributed Workflow with Triana Services, to appear,*Concurrency and Computation: Practice and Experience*, 2005.

[8] Condor Dagman, http://www.cs.wisc.edu/condor/dagman/

[9] I. Curington and M. Coutant. AVS: A flexible interactive distributed environment for scientific visualization applications. Proceedings of *2nd Eurographics Workshop on Scientific Visualization*, 1991.

[10] K.K. Droegemeier, V. Chandrasekar, R. Clark, D. Gannon, S. Graves, E. Joseph, M. Ramamurthy, R. Wilhelmson, K. Brewster, B. Domenico, T. Leyton, V. Morris, D. Murray, P. Plale, R. Ramachandran, D. Reed, J. Rushing, D. Weber, A. Wilson, M. Xue, and S. Yalda. Linked environments for atmospheric discovery (LEAD): A cyberinfrastructure for mesoscale meteorology research and education. Preprints, *20th Conference on Interactive Information Processing Systems for Meteorology, Oceanography, and Hydrology*, Seattle, WA, American Meteorological Society, 2004.

[11] N. Elliott, S. Kohn, and B. Smolinski. Language Interoperability for High-Performance Parallel Scientific Components. *International Symposium on Computing in Object-Oriented Parallel Environments (ISCOPE)*. 1999

[12] I. Foster, C. Kesselman, J. Nick, and S. Tuecke. The Physiology of the Grid: An Open Grid Services Architecture for Distributed Systems Integration, www.globus.org/research/papers/ogsa.pdf

[13] Global Grid Forum, The Open Grid Services Infrastructure Working Group. http://www.gridforum.org/ogsi-wg, 2003.

[14] Globus Alliance, IBM, and HP. Web Service Resource Framework. http://www.globus.org/wsrf. 2004.

[15] GridLab, The GridSphere Portal http://www.gridsphere.org

[16] Grid Service Extensions (GSX). http://www.extreme.indiana.edu/xgws/GSX. 2004

[17] Java Community Process, JSR-168 Portlet Specification. http://www.jcp.org/aboutJava/communityprocess/final/jsr168/

[18] Kepler: A System for Scientific Workflows, http://kepler.ecoinformatics.org/

[19] S. Krishnan and D. Gannon. XCAT3: A Framework for CCA Components as OGSA Services. In *HIPS 2004, 9th International Workshop on High-Level Parallel Programming Models and Supportive Environments*. IEEE Computer Society Press, April 26, 2004.

[20] A. Kropp, C. Leue, and R. Thompson. Web Services for Remote Portlets (WSRP), OASIS http://www.oasis-open.org

[21] A. Mayer, S. McGough, N. Furmento, J. Cohen, M.Gulamali, L. Young, A. Afzal, S.Newhouse, and J. Darlington. ICENI: An Integrated Grid Middleware to support e-Science. In: *Component Models and Systems for Grid Applications*, pp. 109–124, Springer, 2004.

[22] J. Novotny. Developing grid portlets using the GridSphere portal framework, http://www-106.ibm.com/developerworks/grid/library/gr-portlets/

[23] Open Grid Computing Environment (OGCE), http://www.ogce.org.

[24] S.G. Parker and C.R. Johnson. SCIRun: A scientific programming environment for computational steering. In *Supercomputing (SC 95)*. IEEE Press, 1995.

[25] A. Slominski, M. Govindaraju, D. Gannon, and R. Bramley. Design of an XML based Interoperable RMI System : SoapRMI C++/Java 1.1. Proceedings of *IPDPS 2001*.

[26] G. von Laszewski, K. Amin, M. Hategan, N.J. Zaluzec, S. Hampton, and A. Rossi. GridAnt: A Client-Controllable GridWorkflow System, In Proceedings *37th Hawai'i International Conference on System Science*, Jan 5-8, 2004

[27] The Weather Research and Forecasting (WRF) Model. http://www.wrf-model.org/

COMPONENTS FOR HIGH-PERFORMANCE GRID PROGRAMMING IN GRID.IT *

Marco Aldinucci,[1] Sonia Campa,[2] Massimo Coppola,[1]
Marco Danelutto,[2] Domenico Laforenza,[1] Diego Puppin,[1,2]
Luca Scarponi,[2] Marco Vanneschi,[2] and Corrado Zoccolo[2]

(1) – Istituto di Scienza e Tecnologie dell'Informazione, ISTI-CNR, Pisa, Italy

(2) – Dept. of Computer Science, University of Pisa, Italy

{Marco.Aldinucci,Massimo.Coppola,Domenico.Laforenza,Diego.Puppin}@isti.cnr.it

{campa,marcod,scarponi,vannesch,zoccolo}@di.unipi.it

Abstract   This chapter presents the main ideas of the high-performance component-based Grid programming environment of the *Grid.it* project. High-performance components are characterized by a programming model that integrates the concepts of structured parallelism, component interaction, compositionality, and adaptivity. We show that ASSIST, the prototype of parallel programming environment currently under development at our group, is a suitable basis to capture all the desired features of the component model in a flexible and efficient manner. For the sake of interoperability, ASSIST modules or programs are automatically encapsulated in standard frameworks; currently, we are experimenting Web Services and the CORBA Component Model. Grid applications, built as compositions of ASSIST components and possibly other existing (legacy) components, are supported by an innovative Grid Abstract Machine, that includes essential abstractions of standard middleware services and a hierarchical Application Manager (AM). AM supports static allocation and dynamic reallocation of *adaptive* applications according to a performance contract, a reconfiguration strategy, and a performance model.

Keywords: structured parallel programming, programming models, adaptive applications, high performance computing, reconfiguration

*This work has been supported by the Italian MIUR FIRB Grid.it project (RBNE01KNFP) on High-performance Grid Platforms and Tools, and by the MIUR CNR Strategic Project L 499/97-2000 on High-performance Distributed Enabling Platforms.

1.   Introduction

In the context of Grid platforms at various levels of integration [18], a Grid-aware application must be able to deal with heterogeneity and dynamicity in the most effective way (*adaptive applications*), in order to guarantee the specified level of performance in spite of the variety of run-time events causing modifications in resource availability (load unbalancing, node/network faults, administration issues, emergencies, and so on). With respect to traditional platforms, when the Grid is taken into account it is much more important to rely on application development environments and tools that both guarantee high-level programmability, application compositionality, software interoperability and reuse, and, they are able to achieve high-performance and the ability to adapt to the evolution of underlying technologies (networks, nodes, clusters, operating systems, middleware) [8, 17, 21, 26, 27, 30, 32]. Achieving this *high-level view* of Grid application development is the basic goal of our research, in the *Grid.it* national project [20] and in associated initiatives at the national and European level.

In order to be able to design, develop and deploy such kind of high performance Grid-aware applications efficiently, we are interested in innovative programming environments that

i) support the programmers in all the activities related to parallelism exploitation, by providing some kind of *structured* primitives for parallelism exploitation;

ii) allow to achieve full *interoperability* with existing software, both parallel and sequential, either available in source or in object form;

iii) support and enforce reuse of already developed code in other applications.

In particular, we want to exploit the experience of our group in the design and implementation of structured parallel programming environments [13, 22, 23] to target Grids composed of clusters or networks of heterogeneous workstations [1, 3, 5, 6, 15]. We think that a *component-based* programming environment is a suitable starting point to achieve the goals just stated.

In this work, we discuss the essential features of a programming environment that is based both on the component model and on structured parallelism. The programming environment is a layer of a larger picture, such as the one in Table 1. We will first discuss the features of a component-based parallel programming model. Then, we'll take into account how these features are currently or will be soon supported in ASSIST. ASSIST is a structured parallel programming environment that was originally designed to address cluster and networks of TCP/IP workstations only, in the framework of the Italian national project ASI-PQE2000 [31]. We show that ASSIST is a suitable basis to capture all the desired features in a flexible and efficient manner, and, in particular, we discuss how the original ASSIST environment is currently being transformed

Table 1. Layered software architecture for Grid-aware applications

Applications	Complex, multidisciplinary applications. Programmer only expresses the kind of parallelism, without being concerned with any detail involved in its exploitation or related to the fact that the target architecture is a Grid. The user supplies, for each parallel component, a performance contract to be satisfied.
High-performance programming environment	Exploitation of structured parallelism, basic mechanisms supporting compositionality, interoperability and reuse of existing software, support for the dynamic control of performance contracts and adaptivity.
Grid abstract machine	Functionalities and mechanism supporting the programming environment, including dynamic application management and all the needed features from the *resource, collective* and *connectivity* levels of Grid middleware
	Basic hardware-software platform

into a component-based, Grid-aware parallel programming environment. This evolution of the ASSIST environment is being performed in the framework of the Grid.it Italian national project. Grid.it is a 3 year project involving major research institutions in Italy aiming at providing innovative programming methodologies and tools for Grids. The project has a specific work-package, leaded by our group, aimed at designing and implementing a prototype high-performance parallel programming environment for Grids. Component-based ASSIST is the expected, assessed outcome of this work-package.

1.1 Related Work

Several studies recognized that component technology could be leveraged to ease the development of Grid applications. We assume as reference component standards the CORBA Component Model (CCM), because of its clean and rich component model [24–25], and the Web/Grid Services [19], because they are emerging as the standard infrastructure to integrate heterogeneous systems. The Common Component Architecture (CCA) is a prominent standardization effort, aiming at the definition of a high-performance oriented component architecture [10]. We depart from CCA-based approaches like CCaffeine [11] as we explicitly deal with component composition issues (see Section 2 and Section 3.2).

Our approach differs from that of GridCCM [16], as the latter focuses on communication optimization, while our work targets application adaptivity and Grid-awareness in general.

We are closer to the GrADs project [14] with respect to the concept of adaptivity and in some architectural aspects, but we differentiate in the programming model. Our model, being based on structured parallel programming, has the ability *i)* to synthesize from the parallel structure of applications the performance models used to adapt their computation, and *ii)* to control the application configuration at run-time, using a parametric implementation of the parallel programming constructs.

A closely related project to our one is ProActive [7], which extends the Fractal component framework for Java [9] to support parallel, reconfigurable component architectures on the Grid. We share with the Proactive project the departure from flat component models to move toward explicit component composition, the emphasis on run-time adaptivity of component structures, and the exploitation of these hierarchical structures to manage application reconfiguration. We differentiate from that research as we are not limited to Java, we have instead a well defined, language-based separation between sequential program behaviour and parallel coordination, at the intra- and inter- component levels. Our primary goal in improving component interaction is also different, as we want to exploit a broader set of interaction mechanisms than RMI. On the contrary, Proactive primarily exploits parallel and collective RMI abstractions (and their optimizations e.g. by means of futures) to extend the sequential component framework to a parallel, distributed one.

2.   High-Performance Components for Grid-Aware Applications: Computational Model

The basic features needed to implement a component-based, high performance programming environment targeting the Grid include most of those already implemented in currently available component models, such as JavaBeans [29] or the CORBA Component Model [25]. These features are those needed to implement a distributed or a parallel program, that is they are mainly framework features and communication/interaction mechanisms. Concerning the framework mechanisms, we obviously need handy ways of both creating/instantiating and calling components across the different processing elements of the target architecture. We also need mechanisms and features that offer the programmer the possibility of controlling the parallel behavior of the Grid-aware application.

In the *Grid.it* programming environment, we want to provide such mechanisms in the most abstract way possible. In particular, we want to leave the programmer the ability of concentrate on the functional behavior of the application, as well as on the qualitative aspects of parallelism exploitation. That is, we want to relieve the programmer of the responsibility of directly handling all the details related to the quantitative aspects of parallelism exploitation, and all those related to the usage of specific Grid middleware mechanisms.

To understand the features of a high-performance components environment, it can be useful to distinguish three conceptual levels: a) computational model, b) functional and non-functional interfaces and c) support architecture for Grid-aware applications. In this section, we start to deal with the first issue. The other ones are discussed in successive sections, where the Grid.it approach based on ASSIST is presented. From the point of view of the computational model, we propose that high performance components are studied and characterized in terms of the following features: *parallelism, component interaction, composition,* and *adaptivity.*

Parallelism.   In general, components have an internal parallel structure (*intra-component parallelism*). It must be possible to configure several, distinct *versions* of the same component, all versions having the same functional interfaces. Moving from one version to another one could be done by recompiling and/or reloading, in the most simple situations (*static versions*); however, we are interested also in parallel components that are able to change their internal structure/behaviour at run time (*dynamic versions*), depending on functional conditions (e.g. predicates on the computation state) and/or on non-functional conditions (e.g. variations in the achieved performance). In addition to intra-component parallelism, the *inter-component parallelism* is fundamental for high-performance component applications as well.

Component Interaction.   Most component-based frameworks supply a way to declare the public services provided by a component and to invoke a service provided by a (remote) component. The mechanism is based on the uses/provides port abstraction. While being essentially a new edition of the RPC/RMI paradigm, this is sufficient to guarantee proper interactions between components, when they follow this simple client/server model. As an example, task farm computations (that is, embarrassingly parallel ones) can be implemented very efficiently using these mechanisms.

However, different mechanisms are needed to implement other parallel patterns. For instance, pipelines cannot be easily expressed by means of the uses/provides port mechanism. Therefore, we assume that at least two distinct mechanisms are implemented:

- **events,** that is a way to register event handlers and to propagate events through the component network. This mechanism is already present in CCM. It can be exploited to implement all the typical asynchronous activities of parallel computations, such as monitoring.

- **streams,** that is a way to have uses/provides ports that implement data-flow-like channels for sequences of unidirectional typed communications, without incurring in the performance penalties related to the return messages and synchronizations typical of the plain uses/provides port mechanism.

In general, a component has several input streams and several output streams, that can be used in a *data-flow* and/or in a *nondeterministic* fashion. A rich set of interaction mechanisms, which are typical of the parallel computation models, is fundamental also in order to implement higher-level abstractions, such as complex workflow-based PSEs.

Composition.   Currently available component models allow components to interact in several different ways. However, only a few of them (e.g. CCM with the assembly construct) consider component composition as a primitive operation. In our opinion, composition is fundamental to allow more and more complex parallelism exploitation patterns to be developed and provided to the user as components. As derived from our experiences with algorithmic skeletons, as soon as an efficient mechanism to exploit basic parallelism patterns is available (see [4, 12]) then the need for nesting/composition mechanisms arises (see [5, 13]). By exploiting pattern composition, new parallel patterns can be programmed, best suited to user's needs. Furthermore, by properly restricting the visibility of user-defined, composed parallel patterns, different degrees of programmability of parallel applications can be presented to different classes of users.

In our context, we assume to design a *structured*, component-based programming environment, and we actually want to be able to exploit composition of components to provide *new*, non-primitive components supporting the development of Grid-aware parallel applications.

In conclusion, we need to define complex computation structures by means of the parallel composition of parallel components. A composition of components can be defined and reused as a new component in more complex structures. We assume that a general, explicitly parallel structure is encapsulated into a component in order to create a basic parallel component.

Adaptivity.   A *Grid-aware* application must be able to deal with heterogeneity and dynamicity in order to guarantee the specified level of performance, in spite of the variety of run-time events that can change resource availability. A component must be characterized by *non-functional interfaces*, related to the performance control, and by features that allow the programmer to specify how the computation adapts at run time. Moreover, these features have to be implemented efficiently at the run-time support level.

A strong relationship exists between the four features stated above. They must be integrated consistently in a global approach, framework, or better, in a programming model for high performance Grid programming. In Grid.it, we use ASSIST as the programming model able to satisfy this requirement.

3.   ASSIST as the Basic Programming Model for High-Performance Components

We introduce ASSIST as derived from the NOW/COW programming environment as *the* way to denote parallel, high-performance, component-based, Grid applications. We also discuss how already implemented ASSIST features match the requirements emerging from the previous Section discussion or can be exploited/improved to match such requirements.

3.1   Basic Features of the ASSIST Programming Model

ASSIST programs are structured as *generic graphs* (identified by the keyword `generic`), where nodes are parallel or sequential modules and arcs represent typed *streams* of data/objects. No constraint is imposed to the form of graphs, though "structured" graphs, such as those typical of a classical skeleton model, are a notable class of cases that have efficient implementations.

All the interactions that are of interest in the composition of high-performance components are implemented easily and efficiently with the ASSIST streams. Streams are inherently *asynchronous*, however RPC/RMI interactions can be emulated effectively.

Parallel modules are expressed by a *generic skeleton*, called `parmod`. In this context, "generic" means that a `parmod` is a general-purpose construct that can be tailored, for each application, to specific instances of classical stream-parallel and/or data-parallel skeletons, and also to new forms of regular and irregular parallelism.

A `parmod` operates on multiple input streams and multiple output streams. Several distribution and collection strategies are provided for the input and output streams respectively. Moreover, input streams can be controlled in a *data-flow* or in a *nondeterministic* manner. Nondeterminism is important to model several instances of workflow structures, as well as interaction by *events*.

The parallel computation expressed in a `parmod` is decomposed in sequential units assigned to abstract executors called virtual processors (VP). The `parmod` can have an explicitly defined *internal state* for the duration of the computation. This feature is important in many cases, for example in nondeterministic/reactive computations, as well as in many irregular and dynamic computations.

As in any model for structured parallel programming, the `parmod` construct is characterized by the important property that *the implementation model is parametric*. This means that the realization of the run-time support is largely independent of the actual mapping of the virtual processors of `parmod` onto the real processors: this is true as far as it concerns the distribution of functions, the distribution of data, and the communication.

In the same way, an instance of a parmod is characterized by a *performance model*, which is parametric with respect to the actual realization. In the structured parallel programming community, a large amount of performance models have been provided for many stream-parallel and data-parallel skeletons. In ASSIST, we exploit this experience in order to characterize the behaviour of a parmod in terms of the performance it can offer according to the actual mapping. In many cases, i.e. where the parametric behaviour is predictable, the performance model is recognizable at compile-time, while in other, more dynamic cases, the association of a performance model to the computation, expressed by a parmod, requires some annotation by the programmer and/or the knowledge of the past history of the system or application. All the performance models can be made available, in a sort of *Performance Model Repository*, to the strategies implemented by the compiler and to the run-time support.

The "parametricity" feature (parametric implementation model and parametric performance model) is the basis for the implementation of adaptive strategies in high-performance Grid-aware applications. How this issue is dealt with in Grid.it will be discussed in detail in Section 4.1.

The "genericity" of the parmod construct offers an interesting opportunity to express *adaptive parallel computations* by program. That is, the same parmod (the same collection of virtual processors, input and output streams, and state variables) can express different parallelism forms according to the value of the internal state or of the input values. For example, in the Divide & Conquer implementation of the C4.5 algorithm [31], in different phases the Divide module has a data-parallel or a task-farm-like behaviour, in order to optimize the available parallelism in each phase of the computation. Because of the huge amount of data associated to the internal state, this flexible implementation in ASSIST is much more efficient than an equivalent version in which the data-parallel phase and the farm-like phase are expressed by different specific skeletons.

In other words, ASSIST offers a powerful feature for expressing and implementing adaptive computations: the same computation can be expressed according to several alternative strategies. As we describe in Section 4.1, alternative strategies can be associated to values generated by the program at run-time, or they can be selected according to the actual performance values with respect to the performance contract.

ASSIST modules can refer to *external objects* during any phase of the computation, i.e. to objects not defined by the ASSIST coordination language and, consequently, that are referred according to their specific interfaces/APIs. In addition to libraries and system facilities (I/O, file, data bases), current external objects in ASSIST are shared variables, CORBA remote objects, storage objects and data repositories. The existence of external objects is a further facility to express alternative strategies for adaptive computations.

ASSIST provides *full interoperability with CORBA*: not only an ASSIST program can act as a client of a CORBA server, but, most significantly, an ASSIST program can be automatically compiled as a CORBA object with RMI-like synchronous invocations or with stream-like asynchronous data passing. It has been shown [2] that the overhead introduced by the program transformation is definitely acceptable for many parallel applications.

This experience proves that interoperability features can be merged efficiently into the ASSIST model, in order to design applications as composition of components, some of which will eventually be parallel.

3.2   ASSIST and Components

We figured the road-map transforming ASSIST programs/modules to components as follows: first, we allow ASSIST modules to be encapsulated as components in existing, well known component frameworks; then we include in the component framework all the mechanisms needed to implement high-performance applications; eventually, we integrate the framework in such a way that parallel ASSIST components and existing legacy components can coexist in an high-performance parallel Grid application. Following this road-map, our current research [2] will produce the next version of ASSIST (2.0), where an ASSIST program, expressed either as a single ASSIST module or as a graph of parallel or sequential modules, is considered as *a high-performance component* which:

a) *can be composed using standard component frameworks*, in addition to the native ASSIST mechanisms,

b) *exports non-functional interfaces and features automatic support to adaptive applications*, which will be discussed in the next section.

W.r.t. standard frameworks, we are experimenting several solutions based on *Web Services* (WS) and the *CORBA Component Model* (CCM). The implementation approach is similar to the one already adopted for CORBA 2 interoperability, i.e. the compiler generates bridge ASSIST modules, which support the various kind of component ports related to the functional interfaces, as well as ports related to the non functional interfaces.

From the point of view of compositionality, the ASSIST based approach offers the following features:

i) a component (either WS or CCM) encapsulates an ASSIST (sub)graph,

ii) components can be *composed according to ASSIST mechanisms* or according to the mechanisms of the *standard component framework* adopted.

In the first solution of point *ii)*, two or more components, being ASSIST graphs, are composed by the `generic` construct, which describes the structure of the resulting ASSIST graph in terms of component modules and streams (and possibly external objects). The composed ASSIST program is automatically compiled into a standard component. This solution can be adopted when the components are all (new or existing) ASSIST programs whose ASSIST source code is available.

In the second solution of point *ii)*, the programmer uses the interaction mechanisms of the component framework (ports) to compose two or more components. This approach is typically adopted when one or more of the components of the application are existing (legacy) components that have not been designed in ASSIST.

4. Support Architecture for Applications Based on High-Performance Adaptive Components

According to what stated in previous sections, several critical points have to be addressed when we tackle adaptivity control in Grid-aware applications. In particular, non-functional interfaces and reconfiguration strategy and application management have to be taken into account. In this section, we will discuss how these features have been taken into account in the design of our prototype of the component-based parallel programming environment ASSIST 2.0.

4.1 Non-Functional Interfaces and Reconfiguration Strategy

We assume that a Grid-aware application is a composition of high performance components. That is, we restrict to the case where no legacy, non-parallel components are used in the application.

Let us consider the case in which such components are ASSIST components (graphs of sequential modules and/or parmods), composed by means of the generic graph construct. In addition to functional interfaces, that are automatically generated at compile time (out of the ASSIST code), each component is characterized by *non-functional interfaces*. They are expressed as annotations in a proper formalism, which is translated by the compiler into a run-time representation based on XML. Such annotations convey information about *performance contract, reconfiguration strategy,* and *allocation constraints*. The template of an ASSIST component thus assumes the form shown in Figure 1.

Performance Contract.   Many parameters can be used to specify the performance level that is required for the application. In this chapter, we refer to the processing bandwidth (service time) in stream-based computations, and/or to the completion time, which is significant also for non-stream computations. However, the following discussion is largely independent of the specific performance parameters adopted.

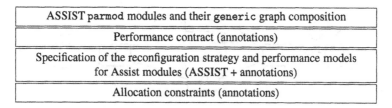

Figure 1. Template of an ASSIST component

The performance contract can be specified for the *whole* application and/or for every *single* component. If the required performance of every component is specified, the required performance of the whole application can be derived at compile time using the knowledge of the graph. For example, the methodology of queueing networks can be used, both in the case of asynchronous stream and in the case of RMI-like interaction. If only the whole application performance is specified, it is still possible to derive rough information at compile and run time on the performance of the single components, using profiling estimates and, most important, taking into account that an ASSIST component is a composition of ASSIST modules for each of which a performance model may be known on the basis of a Performance Model Repository.

Additional information related to communication bandwidth and latency must be estimated; of course, the reliability of this information may not be very accurate. However it is exactly because of these and other inaccuracies, which are inherent of Grid platforms, that we need a support for adaptive Grid-aware applications.

Reconfiguration Strategy. For each component, the application designer specifies which way the component has to be restructured at run-time if and when the performance contract happens to be no longer satisfied.

The reconfiguration strategy is basically expressed in ASSIST with the addition of some annotation. In Section 3.1 we saw that ASSIST allows the programmer to express alternative strategies (e.g. different parametric forms of parallelism) directly in the same program, when their activation depends on the values of some program variables (e.g. the internal state of a parmod). Moreover, the programmer can also specify that alternative strategies must be activated when the performance contract is violated. Let us consider the following example (Figure 2).

a) Component C1 is an interface towards a Grid memory hierarchy, that virtualizes and transforms data sets available on the Grid into two streams of objects, the first one (whose elements have an elementary type) is sent to C2, and the other (whose elements have array type) is sent to C3. C1 may

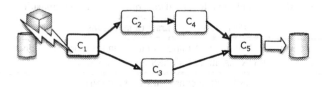

Figure 2.   Example of an adaptive application expressed by parallel components

be an existing component available on the Grid, mediated by an ASSIST program.

b) C2 is a parallel component encapsulating an ASSIST program. The reconfiguration strategy of C2 specifies that

- "by default" C2 is a sequential module executing a certain function F;

- when the current performance level must be adjusted C2 is transformed into a farm computation whose workers execute the same function F. The actual number of workers will be determined *parametrically* at run-time, according to the performance model and to the current availability of resources.

c) C3 is a parallel component encapsulating an ASSIST data-parallel program operating on each stream element of array type. As in the case of C2, the strategy of C3 specifies that its parallelism degree (i.e. the amount of real processors onto which the virtual processors are mapped) can be modified, if needed.

d) C4 is a parallel component encapsulating an ASSIST program which, by default, is a sequential module, but it can be restructured into a parmod operating on the input stream according to a data-parallel or to a task farm style, *depending* on the values of the parmod state and on the input values themselves. Therefore, in this case the adaptation principle is applied at two levels: at the program level and at the run-time support level.

e) C5 is a parallel component encapsulating an ASSIST program operating nondeterministically on the input values received from C3 or C4, and transforming the two streams into a data set.

Let us assume that, at a certain time, some monitoring activity signals that C2 is becoming a bottleneck and that this causes a substantial degradation of performance of the whole application. C2 can be transformed into a version with suitable parallelism degree. In this case other components may have to be restructured (e.g. C4,C5) in order to guarantee the level of performance. As previously stated, this is possible according to a global strategy based on the knowledge of the application structure.

When restructuring data-parallel components (C3), the strategy must be applied also to the re-distribution of the data constituting the internal state of a parmod.

More sophisticated strategies can be expressed in ASSIST than those shown in the example: the strategy of C4 could depend just on performance requirements instead of predicates on the internal state, and other alternative strategies could exploit external objects, as opposed to strategies based on the stream composition only.

Allocation Constraints. In general, restructuring high-performance components involves resources belonging to different Grid nodes. In the example above, the new workers of C2 can be allocated onto processors of a cluster from a different Grid node. There are instead several cases in which we must put constraints on resource allocation. For instance, several components (C1 and C5, say) can be executed only on certain Grid nodes and no reconfiguration is permitted, either due to security or privacy reasons, or to requirements related to the kind of resources needed to operate on the data sets. This kind of information has to be associated with the reconfiguration strategy of every component.

4.2 Application Management for Reconfiguration

We eventually come to the point where the implementation of the high-performance component framework has to be taken into account. The software architecture of Grid.it component-based parallel programming environment is organized as shown in Figure 3. The run-time environment of ASSIST 2.0 is implemented on top of a *Grid Abstract Machine* (GAM), which in turn is

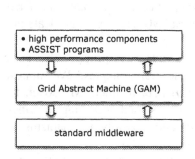

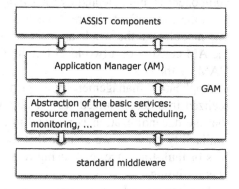

Figure 3. Grid.it software architecture

Figure 4. Grid Abstract Machine

implemented on top of existing middleware (currently a version of the Globus Toolkit) and realizes the functionality needed by the programming environment to support high-performance, component-based Grid-aware applications.

As shown in Figure 4, the GAM consists of the Application Manager and of the *abstraction of the services* for Resource Management and Scheduling, Monitoring, and other services (accounting and so on).

Application Manager Structure.   The Application Manager (AM) has a *hierarchical structure*. Figure 5 illustrates the simple case of an application consisting of just one component, structured as a graph of ASSIST modules.

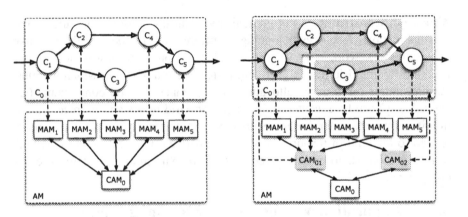

Figure 5.   Example of an ASSIST component

Figure 6.   Two interacting ASSIST components

Each ASSIST module is associated with a *Module Application Manager* (MAM): each MAM_i is responsible of the configuration control of the associated module. A global strategy for the configuration control of the whole component is implemented by the *Component Application Manager* (CAM).

In the example of Figure 2, each of the components $C_1, \ldots C_5$ consists of one module and the whole application is wrapped as a component C_0. Then, the AM consists of a two-level tree in which $MAM_1, \ldots MAM_5$ are leafs and CAM_0 is the root.

Application management is, in principle, a centralized process. It can be realized in a decentralized way, according to several strategies. We suppose that the decentralization is realized in a *hierarchical* manner. Moreover, for availability reasons, we assume that the root is designed according to principles of fault-tolerance, e.g. using redundancy and, possibly, mechanisms for checkpointing.

The hierarchical structure can be extended at any level according to the compositionality and abstraction strategy adopted for the application. For example,

Figure 6 shows the same application of Figure 5 in which we recognize two components, C_{01} consisting of modules C_1, C_2 and C_4, and C_{02} consisting of modules C_3 and C_5. The whole application, which can be seen as the composition of C_{01} and C_{02}, is considered as a component C_0. Thus, in addition to the same leaf managers ($MAM_1, \ldots MAM_5$), we have CAM_{01} and CAM_{02} at the second level, and CAM_0 at the root.

Module Application Managers (MAMs). The MAM level is an ASSIST abstraction, independently of the fact that the application is structured as a (hierarchical) composition of higher-level components. At the MAM level we implement the configuration control of the single ASSIST modules (parmods) exploiting the associated performance model. As introduced in Section 3, a *Performance Model Repository* is provided in the programming environment and it is updated according to the history of the applications running in the system. The specific performance model of each module, to be found in the performance model repository, can be recognized

- by the compiler, according to the knowledge of some parallelism forms. Examples of parallelism forms which are statically recognizable in ASSIST are farms (with and without state), data parallel computations with fixed or variable static stencils, and some mixed combinations of stream- and data-parallelism;

- by the programmer, in all the cases in which his knowledge is more accurate, and/or a new parallelism form and the associated performance model are expressed by properly and specifically instantiating a parmod construct.

As discussed in Section 3 and 4, the reconfiguration strategies of an ASSIST module can exploit different forms of parallelism, different data distribution/-collection strategies, and the usage of external objects. Moreover, in case of data-parallel behaviour, data can be redistributed at run-time according to *load balancing* strategies that cannot be (have not been) recognized at compile-time. Notice that, for stream-parallel farm-like structures, load balancing is always implemented by the run-time support .

MAM behaviour is basically *event-driven*, where the events indicate the opportunity/necessity for restructuring the associated ASSIST module. One kind of event is generated according to the outcome of Monitoring service invocations. In this case, the MAM can provide the following sequence of actions:

- a restructuring strategy is taken into account, either based on the ASSIST alternative descriptions or on load balancing for data parallel modules;

- in case of alternative parallel strategies, the performance model from the performance model repository is applied, a proposed solution to reconfiguration is derived,

- the non transient nature of the event is assessed and therefore

- the father CAM is informed about this proposal.

The MAM can also receive an event by the father CAM indicating that it has to apply a restructuring strategy because a global variation of performance has been detected. For example, in the computation of Figure 6, CAM_{01} can ask MAM_4 to apply a reconfiguration strategy in order to increase the C_4 bandwidth in consequence of an increase of bandwidth of C_2.

Component Application Managers (CAMs).   Each CAM applies control strategies *at a global level* for the associated component.

As indicated above, a CAM can receive proposals of restructuring by the child MAMs. In this case, the CAM has to apply a global performance model (e.g. queueing network based model) in order to individuate a good solution to the restructuring of one or more of the children modules. The Allocation Constraints, indicated in the non-functional interfaces of the component, are also applied during this process.

Recursively, a CAM can receive reconfiguration requests from father CAMs, and can send them reconfiguration proposals. The root CAM (CAM_0) is responsible for the final decisions in the global reconfiguration control which, as seen, is a sort of parallel and asynchronous Divide & Conquer strategy applied along the hierarchical Application Manager structure.

At each CAM level, the Resource Management and Scheduling services provided by the Grid Abstract machine are utilized. Notice that such services do not necessarily coincide with the services in the standard middleware, instead they represent the abstraction that are strictly needed by the Application Manager. That is, a "RISC-like" GAM is defined, though starting from a monolithic Middleware level; in the next future, this GAM service structure could be the basis for the proposal of a new Risc-like Middleware level.

5.   Experiments

The features to be included in the Grid Abstract Machine have been experimented using Lithium, a full Java, RMI based, structured parallel programming environment [3]. Lithium has been often used to experiment solutions that have been then moved to ASSIST, as the former is much more compact and easy to modify than the latter. These experiments showed that

- almost perfect scalability can be achieved, even in the case when heterogeneous resources are used for the execution. The measured execution times are usually no more than 5% away from the ideal ones.

- good tolerance to typical "dynamic" situations, such as the presence of faulty nodes, can be achieved. In presence of a number of faulty nodes not exceeding 20% of the nodes used to compute the parallel application, an increase of less than 10% of the total execution time has been measured.

The proposed AM organization and behaviour, described in Section 4, have then been evaluated on some ASSIST examples, emulating the dynamic features of the run-time support and of the MAM/CAM hierarchical organization. The implementation of a first version of this support is on-going. Figure 7 shows the results achieved in a set of reconfiguration experiments [28]. The experiments have been performed using an application whose structure was a pipeline of three stages: the first and the third stages are data servers and stream managers,

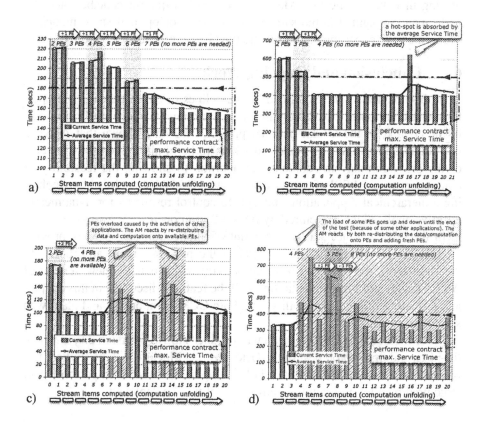

Figure 7. Experiments in dynamic restructuring of parallel components

and the second stage is a data parallel version of the finite difference method for solving differential equations (Jacobi method).

Figure 7-a shows how the Application Manager can satisfy the performance contract (service time) by increasing the amount of real processors onto which virtual processors of a data-parallel stencil computation are mapped. Figure 7-b shows the effect of transient variations of system load, that have no effect in the performance in the long period (we employed an exponential mean, reset at every reconfiguration). Figure 7-c shows the more serious effects of a perturbing overload caused by the creation of a new application onto the same processing nodes. In this case, we assume that no more processing nodes are available, thus only the load balancing solution is attempted, with a suitable redistribution of data partitions implemented directly by the run-time support. Fig. 7-d shows a situation similar to Fig. 7-c: the difference is that more processing nodes are now available, and, after a first attempt of applying data redistribution, the degree of parallelism of the data parallel module is successfully increased.

These experiments have been performed on a heterogeneous cluster, confirming that the Application Manager overhead is quite acceptable, and the performance obtained is the same as in the case of optimal static mapping. More intensive experiments to evaluate the Grid overhead are on-going.

6.   Conclusion and Ongoing Work

In this chapter we have outlined the guidelines of our research in high performance component-based programming environments which are Grid-aware, in the context of the Grid.it national project. We have shown that ASSIST is a suitable programming model on which to build all the complex features of the programming environment. In addition to showing the feasibility of component-based ASSIST, we have proposed a Grid Abstract Machine, including a hierarchical Application Manager to control resources for dynamically adaptive applications, structured by ASSIST components.

In the Grid.it project, a large amount of application case-studies provide intensive experiments and benchmarks of the proposed ideas and tools. In the short term, the on going research activity will produce a new version of ASSIST with a full implementation of all the features discussed in this chapter and providing full interoperability with both CCM components and plain web services as well.

In the medium term, the research will produce ASSIST version 2.0, in which the ideas and first prototypes for the Grid Abstract Machine will be studied, implemented and evaluated and the whole, component-based, high-performance, structured parallel programming environment will be deployed.

References

[1] M. Aldinucci, S. Campa, P. Ciullo, M. Coppola, S. Magini, P. Pesciullesi, L. Potiti, R. Ravazzolo, M. Torquati, M. Vanneschi, and C. Zoccolo. The implementation of AS-SIST, an Environment for Parallel and Distributed Programming. In *Euro-Par 2003 Parallel Processing*, LNCS, 2790:712–721, Springer, August 2003.

[2] M. Aldinucci, M. Coppola, M. Danelutto, M. Vanneschi, and C. Zoccolo. ASSIST as a Research Framework for High-performance Grid Programming Environments. Technical Report TR-04-09, Dept. Computer Science - University of Pisa, February 2004. TR-04-09 available at http://www.di.unipi.it/Ricerca/TR.

[3] M. Aldinucci, M. Danelutto, and P. Teti. An advanced environment supporting structured parallel programming in Java. *Future Generation Computer Systems*, 19(5):611–626, 2003.

[4] P. Au, J. Darlington, M. Ghanem, Y. Guo, H.W. To, and J. Yang. Co-ordinating heterogeneous parallel computation. In L. Bouge, P. Fraigniaud, A. Mignotte, and Y. Robert, editors, *Europar '96*, pages 601–614. Springer, 1996.

[5] B. Bacci, M. Danelutto, S. Orlando, S. Pelagatti, and M. Vanneschi. P^{3}L: A Structured High level programming language and its structured support. *Concurrency Practice and Experience*, 7(3):225–255, May 1995.

[6] B. Bacci, M. Danelutto, S. Pelagatti, and M. Vanneschi. SkIE: a heterogeneous environment for HPC applications. *Parallel Computing*, 25:1827–1852, December 1999.

[7] F. Baude, D. Caromel, and M. Morel. On hierarchical, parallel and distributed components for Grid programming. In: *Component Models and Systems for Grid Applications*, pp. 97–108, Springer, 2004.

[8] F. Berman, R. Wolski, H. Casanova, et al. Adaptive Computing on the Grid using AppLeS. *IEEE Trans. On Parallel and Distributed Systems*, 14(5), 2003.

[9] E. Bruneton, T. Coupaye, and J. B. Stefani. Recursive and Dynamic Software Composition with Sharing. In *7th International Workshop on Component-Oriented Programming (WCOP02)*, ECOOP 2002, Malaga, Spain, June 2002.

[10] CCA working group. The common component architecture technical specification - version 0.5 (as amended through 3/5/2001). http://www.cca-forum.org.

[11] Ccaffeine home page, 2003. http://www.cca-forum.org/ccafe/.

[12] M. Cole. *Algorithmic Skeletons: Structured Management of Parallel Computations*. Research Monographs in Parallel and Distributed Computing. Pitman, 1989.

[13] M. Cole. Bringing skeletons out of the closet: a pragmatic manifesto for skeletal parallel programming, *Parallel Computing*, 30(3):389–406, 2004.

[14] H. Dail, O. Sievert, F.Berman, H. Casanova, A. YarKhan, S. Vadhiyar, J. Dongarra, C. Liu, L. Yang, D. Angulo, and I. Foster. Scheduling in the Grid Application Development Software Project. In: *Grid Resource Management: State of the Art and Future Trends*, pp. 73–98, Kluwer, 2003.

[15] M. Danelutto, R. Di Meglio, S. Orlando, S. Pelagatti, and M. Vanneschi. A methodology for the development and support of massively parallel programs. *Future Generation Computer Systems*, 8(1–3):205–220, July 1992.

[16] A. Denis, C. Pérez, T. Priol, and A. Ribes. Bringing High Performance to the CORBA Component Model. In *SIAM Conference on Parallel Processing for Scientific Computing*, February 2004.

[17] B. Ensink, J. Stanley, and V. Adve. Program Control Language: a programming language for adaptive distributed applications. *Journal of Parallel and Distributed Computing*, 63(11):1082–1104, 2003.

[18] I. Foster and C. Kesselman, editors. *The Grid: Blueprint for a New Computing Infrastructure*, chapter 11, The Globus toolkit. Morgan Kaufmann Pub., S.Francisco, CA, 1998.

[19] I. Foster, C. Kesselman, J.M. Nick, and S. Tuecke. Grid Services for Distributed System Integration. *Computer*, 35(6):37–46, June 2002.

[20] Grid.it project: Enabling platforms for high-performance computational grid oriented to scalable virtual organizations. MIUR, FIRB National Research Programme, November 2002, http://www.grid.it.

[21] K. Kennedy, M. Mazina, J. Mellor-Crummey, K. Cooper, L. Torczon, F. Berman, A. Chien, H. Dail, O. Sievert, D. Angulo, I. Foster, D. Gannon, L. Johnsson, C. Kesselman, R. Aydt, D. Reed, J. Dongarra, S. Vadhiyar, and R. Wolski. Toward a Framework for Preparing and Executing Adaptive Grid Programs. In *Proc. of NSF Next Generation Systems Program Workshop (IPDPS 2002)*, 2002.

[22] H. Kuchen. A Skeleton Library. In *Euro-Par 2002, Parallel Processing*, LNCS, 2400:620–629, Springer, August 2002.

[23] S. MacDonald, J. Anvik, S. Bromling, J. Schaeffer, D. Szafron, and K. Taa. From Patterns to Frameworks to Parallel Programs. *Parallel Computing*, 28(12):1663–1684, December 2002.

[24] Object Management Group. The Common Object Request Broker: Architecture and Specification, Minor revision 2.4.1, http://www.omg.org, 2000.

[25] Object Management Group. Corba Component Model, v3.0, November 2001. Document ptc/2001-11-03, available at http://www.omg.org/.

[26] C. Pérez, T. Priol, and A. Ribes. PaCO++: a parallel object model for high performance distributed systems. In *Distributed Object and Component-based Software Systems, Hawaii Int. Conf. System Sciences*, IEEE Comp. Soc., 2004.

[27] T. Priol. Programming the Grid with Distributed Objects. In *Proc. of Workshop on Performance Analysis and Distributed Computing (PACD 2002)*, 2002.

[28] L. Scarponi. Supporti alla programmazione Grid-aware – Esperienze di allocazione dinamica di programmi ASSIST (Grid-aware programming support: experiments in dynamic program allocation with ASSIST). Master's thesis, University of Pisa, April 2004, (in Italian).

[29] Sun. Javabeans home page. http://java.sun.com/products/javabeans, 2003.

[30] D. Thain, T. Tannenbaum, and M. Livny. *Grid Computing: Making the Global Infrastructure a Reality*, chapter Condor and the Grid. Wiley, 2003.

[31] M. Vanneschi. The programming model of ASSIST, an environment for parallel and distributed portable applications. *Parallel Computing*, 28(12):1709–1732, December 2002.

[32] R. van Nieuwpoort, J. Maassen, R. Hofman, T. Kielmann, and Henri Bal. Ibis: an Efficient Java-based Grid Programming Environment. In *ACM JavaGrande ISCOPE 2002 Conference*, pp. 18–27, Seattle, WA, November 2002.

TOWARDS BUILDING A GENERIC
GRID SERVICES PLATFORM:
A COMPONENT-ORIENTED APPROACH

Jeyarajan Thiyagalingam, Stavros Isaiadis, and Vladimir Getov
Harrow School of Computer Science
University of Westminster
Harrow, London, U.K.
T.Jeyan@westminster.ac.uk
S.Isaiadis@westminster.ac.uk
V.S.Getov@westminster.ac.uk

Abstract   Grid applications using modern Grid infrastructures benefit from a rich variety of features, because they are designed with built-in exhaustive set of functions. As a result, the notion of a lightweight platform has not been addressed properly yet, and current systems cannot be transplanted, adopted or adapted easily. With the promise of the Grid to be pervasive, it is time to re-think the design methodology for next generation Grid infrastructures. Instead of building the underlying platform with an exhaustive rich set of features, in this chapter, we describe an alternative strategy following a component-oriented approach. Having a lightweight reconfigurable and expandable core platform is the key to our design. We identify and describe the very minimal and essential features that a modern Grid system should always offer and then provide any other functions as pluggable components. These pluggable components can be brought on-line whenever necessary as demanded implicitly by the application. With the support of adaptiveness, we see our approach as a solution towards a flexible dynamically reconfigurable Grid platform.

Keywords: generic Grid platform, lightweight Grid platform, adaptive Grid, adaptive Grid service

1.   Introduction

In recent years, significant efforts have been made towards designing and building advanced Grid infrastructures. One of the main priorities in building new Grid systems is to assure longevity and flexibility. In order to support these two seamlessly, the underlying Grid platform is normally built with a rich set of features, such that the requirements of any Grid application need a subset of the complete list provided by the platform. Recent standardisation efforts and software for Grids [13–14] also aim at providing infrastructures with all possible features built-in. The Open Grid Services Architecture (OGSA) [10], on which most of the current Grid implementations are based, is built as a feature rich specification. This approach ensures that any service request from applications is covered by the complete set of features offered by the platform.

Complexity (in terms of interactions, manageability and maintainability) of the implementation of any Grid platform based on this philosophy will be very significant. For example, upgrading a service component in this model is a difficult task. When one service component is modified, other components also need to be modified. Further, deployment of these Grid systems demand considerable computing resources. Managed and/or un-managed migration of these Grid platforms is also a challenging task. Nor can they be extended very easily in terms of functionalities and capabilities. For example, layering an existing Grid platform on a lab of machines involve considerable effort in configuring. Difficulties in configuring the platform involves removing or disabling the unnecessary features and in extending the system capabilities. In summary, current Grid systems are failing to address the issue of generality and reconfigurability. This is not a design flaw; instead they are designed with exhaustive set of services targeting longevity and flexibility – resulting in highly complex platform, impeding the expandability.

This chapter summarises the current status of our ongoing work on designing the architecture of a generic Grid platform. We identify a generic set of features that should be common to any Grid system, while addressing the issue of longevity and flexibility. Further, we also address the issue of a "lightweight platform". The motivation in identifying this common set of core features is to standardise the road map for development of future Grid systems, which should be adaptive and intelligent while retaining the features of flexibility, longevity and expandability. The overall contributions of this chapter are:

- Proposing a generic Grid platform with minimal complexity but with core essential features;

- Providing a seamless way of extending the platform capabilities by means of component introduction;

- Means to offer and guarantee more flexibility to the end users.

This chapter is organised as follows: Section 2 provides the background for the chapter. Section 3 identifies the common set of features found across different Grid platforms. Section 4 discusses the architectural aspects of this core Grid platform while Section 5 concludes the chapter with future works.

2. Background and Related Work

The original motivation behind the OGSA development [10] was to offer ubiquitous support for Grid infrastructures by converging Web Services and Grid Services. The Open grid Services Infrastructure (OGSI) specification [17], on which OGSA relies, included necessary extensions to support stateful web services. However, the fact that these extensions were heavily object oriented and the interoperability issues with the Web Services and XML, impeded the adoption by the Grid Community.

Refactoring OGSI led to consider the Web Services Resource Framework (WSRF) [8–9], which constitutes specifications for different web services and management services. WSRF retains all OGSI functions but all these are enhanced to meet the web services specification, for example WS-Addressing [5]. The idea of adaptivity in Grid Systems has been discussed in [4] where main emphasis is given either at the very low level, the middleware level, or at the application level. However, the idea of service level adaptivity for heavily componentised Grid systems has not been addressed in these works.

Reconfigurability at software components level, especially in the context of Grid systems, has not been addressed in the literature. The notion of Web Services is included in our proposed Generic Grid Services Platform both at the higher level and at the lower level. In other words, the platform offers the Grid Service as a web service. Further, componentised functionalities can also be represented as web services. However, the lower level of service interaction is transparent to the end-user or applications. These web-service components are adaptive and an extra layer of flexibility is provided by permitting these components to be re-wired as necessary to provide the reconfigurability.

3. Generic Services

OGSA was derived from use cases of e-business and e-science applications [11]. These applications require more functionality in addition to the fact of being computationally demanding. This has influenced the architectural aspects of OGSA and resulted in functionally-rich and thick platform specification. To identify the minimal set of core features, an equal emphasis must be given to small-scale applications and devices as well, contrary to the approach that OGSA has taken.

The idea of the component-oriented design approach is to componentise the functionality of the set of core features that should be offered by the platform.

Later, the functionality of the platform can be extended by plugging in additional components on-demand. Enabling the generic platform to secure the foreknowledge on these pluggable components, permits the platform to extend the capability as necessary. Further, the platform should also be pro-active when components are introduced in order to inter-relate the operations of different components. For example, the platform should be able to recognise additional operations when a self-healing functionality is plugged in, so that any further negotiations with the fault-tolerant component can be done effectively.

A permanent component implementing a core feature for the generic Grid platform is defined as **Feature**. The information related to an optional component, which might be plugged in whenever its implemented feature is necessary, is defined as **Feature Knowledge**. *Feature Knowledge* related to a specific component provides information only about the component, expected interface, and interaction map across components, which enables cross-component operations. The *Feature Knowledge Set* is a collection of *Feature Knowledge*. Members of the *Feature Knowledge Set* do not implement any of the functionalities. Instead, functionalities are separately and exclusively implemented inside the respective components. In other words, the *Feature Knowledge Set* is the foreknowledge of the engine about pluggable components.

With these definitions, the idea is to design the generic Grid platform with a minimal and essential *Feature Set* and with the necessary *Feature Knowledge Set*. The platform has to be engineered such that new *Feature Knowledge* can be added later on. However, in order to configure a functional Grid platform, it may be necessary to select different permanent components depending on the use case scenario. For example, it is essential to include a resource management component to the generic Grid platform to realise a fully working Grid platform. The reason why it is being added through the knowledge set is to enable the development of tailored components. It is possible to include the resource management inside the generic Grid platform as a permanent component, but such a resource manager should be rich in features and some of them may not be used at all. Consider a use case of a computer laboratory with PCs turning to a Grid system during the middle of the night. The resource management functionality for such a system is completely different than the one for a supercomputer centre.

3.1   Feature Set

The following set of features must be available as part of the core of the proposed Grid platform. These components and their interactions within a Grid system are shown in Figure 1.

- **Core Operating Support**: This results from feature extraction from the Native Platform Services and Transport Mechanisms and OGSA Hosting

Environment from OGSA specification. This feature forms the concrete resource-hosting environment. However, a main difference is that the approach taken in building this layer is similar to building the Java Virtual Machine [16], building the core support for underlying operating system/hardware pair. Once they are in place this feature enables the system to handle the hosting of resources specific to the supported operating systems or hardware components, and the native resource managers manage them. Effectively this feature provides the basic operating skeleton and a hosting environment – an essential feature for a Grid Platform.

- **Core Connectivity Services**: (The connectivity services can also be the part of core operating support) Core connectivity services are to offer networking and transport functions for data transfer across multiple Generic Grid platforms and within the Grid domain. By default, it uses the platform specific connectivity/network/transport parameters (such as protocols) but can be varied by *Feature Knowledge*.

- **Knowledge Engine**: This part interprets and understands the knowledge sets discussed in the next section. This also permits addition of new knowledge sets.

- **Component Management Engine (CME)**: This part manages the different components and triggers actions wherever applicable to handle the cross-component interaction.

- **Service Management Engine (SME) / Service Manager (SM)**: All service operations are orchestrated and coordinated by this kernel. It is also responsible to direct the CME.

3.2   Feature Knowledge Set

As outlined above, effective operation of a Grid system inherently depends on multiple capabilities of the Grid platform, which we decided to componentise. An application, such as the one described in [11] may require introduction and interaction of multiple components for the operation. A careful inspection of [10–11] reveals that the following set of functionalities must be available as separate components so that, whenever necessary, any component providing a required functionality can be brought on line.

3.2.1   Basic Functionality Extension Components.

- Resource Discovery

    When a new resource enters a Grid environment must let the rest of the Grid know what type of services it provides and also to find other available

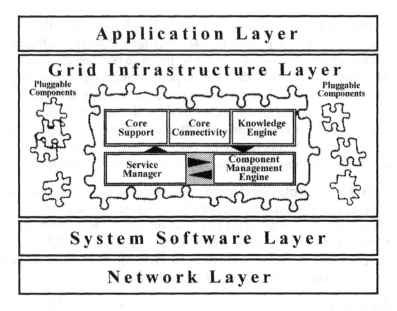

Figure 1.   Simplified layered diagram focusing on core features to be implemented inside the proposed generic Grid services platform.

services in the Grid. Mechanisms have to be provided to support such a dynamic resource discovery scheme. This is usually achieved with the use of a registry along with relevant registration and query mechanisms.

- Accounting / Metering and Pricing of Services / Resources

  These services meter the usage of the resources, while for commercial Grids a pricing/billing component should also be in place to control resource utilisation — (perhaps based on price limits) producing pricing reports and necessary bills. Logging mechanisms are also required for the provision of more advanced services like for example forecasting which makes use of resource usage logs.

- Monitoring

  In a complicated and dynamic Grid environment, Monitoring services for applications, resources and usages can assist in maintaining an "environment", providing valuable information for troubleshooting in case of failures, supplying data regarding user applications and resource usage among other information.

- Data Management

  Data management techniques such as data deployment/migration, data replication and data sharing are common in a Grid environment and should be supported by specialised components. Data migration (or deployment), sharing and replication are important techniques and are sometimes used to support failure/disaster recovery, higher performance through parallel data processing, service continuation through data mirroring (replication), job scheduling and work load balancing and many other procedures.

- Notification / Reporting / Messaging

Notifications and messages are very important in emergency situations like component or resource failures, but can also help in troubleshooting and prevention of unwanted conditions like heavily loaded resources/services or data inconsistencies.

- Virtual Organisations (VO) Management

In terms of available resources, VOs can contribute to the deployment of more scalable and richer Grids. Mechanisms have to be provided that achieve automatic, dynamic VO creation (by merging collaborative networking environments) and VO management.

- Component-based Policy Management and Application

Policies play a very important role in any Grid environment and can be present in almost every aspect of a Grid: resource management, security, accounting, pricing and data management just to name a few. Components/mechanisms that enforce the application of all these policies in an automated manner can be provided here.

3.2.2 Security.

- Authentication / Authorisation and Accounting

The most fundamental notions of security in a distributed environment are those of authentication and authorisation. Authentication requires both the consumer of a service and the service to authenticate themselves to each other. This can be achieved with the use of a Public Key Infrastructure for example. Authorisation controls who has access to which resources and can be provided by simple access lists or more sophisticated techniques.

- Certification

Every resource needs to present a certificate in order to register to the Grid. This certificate can be acquired from an independent Certificate authority and provides such information about the resource as the type of service provided, owner of the resource and other.

- Encryption / Decryption

  A cryptographic infrastructure is important in order to maintain confidentiality of sensitive messages. This is usually achieved using a Public Key Infrastructure to support digital signatures and encryption/decryption of messages.

- Various Security Infrastructure-supporting Components

  The nature of the Grid assumes that many companies, organizations or individuals will participate in a Grid environment. Each of these parties will probably make use of different security infrastructures and techniques. Support services have to be in place to ensure secure interoperability across these different platforms. Such services should minimally include single sign-on, delegation of credentials and intrusion prevention and detection.

- Secure Inter-Grid Communications

  Different Grid platforms should be able to communicate with each other in order to utilise available resources and scale. We have to provide mechanisms to support inter-Grid collaboration and interoperability without compromising security (mainly) or functionality.

3.2.3   Resource Management.

- Provisioning

  Components providing services such as scheduling of resources, reservation and termination are included in this category. Advance reservation, scheduling and provisioning as well as termination mechanisms provide the necessary support for the smooth and efficient utilisation of the available resources. Complimentary services like deadlock resolve mechanisms or freeing resources bound to processes that terminated abnormally, can enhance functionality and increase availability.

- Load Management / Balancing

  Such services can increase performance and resource availability by eliminating possible communication bottlenecks and redistributing workloads of heavily loaded resources to ensure that all resources are used uniformly. Also, load balancing components can ensure that certain requirements are met (or at least at the highest possible degree) by reallocating resources depending on the Grid user demands. For example more resources could be provisioned for a critical or highly prioritised application to ensure increased performance.

■ Scavenging

Most workstation nodes present in a networking environment will remain idle most of the time according to many recent studies. In a Grid environment, utilising these idle resources is of great importance. These resources can be combined to create a huge secondary storage, memory or CPU pool that could substantially improve performance in demanding Grid applications. Scavenging mechanisms, however, should manage idle resources very delicately since the end user response times should not become unacceptable when the user decides to use his machine again.

3.2.4   Added Services.

■ Fault Tolerance

Fault tolerance requires mechanisms for fail-over, workload redistribution, service continuation and notification of other relevant services like self-healing and disaster recovery. Fault tolerance is of extreme importance to real time environments or critical applications where even the lowest possible percentage of down-time might be unacceptable and/or disastrous.

■ Disaster Recovery

Disaster recovery mechanisms are also important in sensitive Grid environments and should ensure continuation of at least the most vital services. They should also take actions to restore system operation and service, resource and application states as soon as possible (perhaps using backup data, previous checkpoints and last known state information).

■ Self-Healing

Self-healing is the ability of a system to monitor its resources, detect failures and plan and apply necessary changes to ensure resource availability and service continuation. Human intervention should be kept to a minimum level and all operations should be performed automatically, with human administrators only being notified in emergency or unresolved situations.

■ Forecasting / Prediction

Forecasting components cooperate with and may require the presence of scavenging, scheduling, workload balancing, metering and logging mechanisms. They can then extract valuable usage pattern information that can be used to predict the amount of time an idle workstation will remain idle and assist that way in job scheduling and workload balancing and reduce execution time costs.

- Optimisation

    In addition to providing services, the platform should also optimise various operational aspects of the system, applications running on them and the interaction of different components in order to provide smooth and efficient operation.

4.   Engineering the Generic Grid Platform

In this section, we use two operational examples to illustrate the operation of the generic Grid platform and then we discuss the design aspects of the platform.

4.1   Functionality

Although the core functionality built inside the generic Grid platform is very minimal, when engaged in supporting an application, the platform must offer all necessary services as required by the application. If such services are not available from the platform itself (either as part of the core-feature set, or as pluggable components), the internal mechanism may decide to secure a specific service from a remote site because of the limited local resources or because it is more efficient to act as a client rather than download and plug-in this particular service component. If the remote site decides to not allow the component to migrate, the platform may react according to a pre-configured policy or may act adaptively. However, submission of a clear job description along with the job, is an essential part of the whole process, as in [13].

The overall operation of the generic Grid platform solely depends on the capability of the SME to accommodate, anticipate and to reconfigure the components plugged in. Successful engineering of such a platform requires clear understanding of the operation of the proposed platform. Here, we consider two different applications with differing requirements to illustrate the operation of the platform.

4.1.1   Operational Example 1.   Consider the third use-case example as described in [11], where a severe storm prediction is considered. Functionally, the following sequence of operations will take place in our proposed platform:

1 The generic Grid platform announces the service availability through UDDI [7].

2 An interested client forwards the job description request to the platform.

3 The SM/CME part of the platform analyses the job description. The job description need to state all the requirements of features and should supply a handle to any proprietary features. These proprietary features can replace an already existing feature in the Grid platform, or can be

a completely new feature. It also verifies whether the set of features requested by the application is not empty. This means that, a request for a feature should exist on either the platform side or the client side. If the platform fails to secure any service (either locally or remotely) the request is terminated with a negative acknowledgement and positively otherwise.

4 Having accepted the job, the SM/CME should authenticate the user and authorise the job for further manipulation. This requires a feature which is not present in the generic Grid platform, but available through on-demand loading (in case of components) or through intra-Grid web-service (in case of web-service). SM/CME identifies the end-point where the "AAA" service is available and forwards the request and obtains the response before proceeding further.

5 The policy component needs to be loaded and consulted to determine the operational policy, if there are any, to be applied.

6 Once the job is authenticated and authorised, the SM/CME builds a list of resources and features and builds the interaction and dependency map between components or services.

7 This in turn requires a consultation with the resource reservation. This necessitates the launching of resource management service which along with the discovery, brokers the requests.

8 In case of reservation is favoured, the SM/CME updates the job status and continues further with the other operations that does not break the dependency or it waits until it is indicated that the resources are available.

9 If the advance reservation was done, then the SM/CME also delegates the provisioning and management tasks to the resource management component.

10 The SM/CME is notified about the resource availability.

11 If the customer wants to monitor the progress, the SM/CME loads and launches the Monitor (Application Part) component, which reports the status of the job at various time steps. This also triggers the launching of Reporting (or Notification or Messaging component) and logging components.

12 SM/CME also launches the pricing/metering component.

13 As this application requests complete reliable service, the self-healing, disaster recovery and fault tolerant components also loaded for this application.

14 Application also involves access to the large databases and this mandates the loading of data management component.

15 High Performance and fair pricing strategy requires application and workload to be distributed evenly and to be balanced as much as possible across the machines. This requires launching "load management and balancing" component.

16 With an application that can spawn multiple other applications, it is necessary to have proper synchronisation. This requires efficient orchestration of the tasks executed on the Grid - and thus launching "Work-Flow" component is inevitable.

17 The actual task servicing begins.

18 Along the time line, the SM/CME also should launch scavenging and resource optimiser components to guarantee the all free cycles are harnessed.

19 Execution Terminates, modules are unloaded one by one, and the result is forwarded to the customer.

4.1.2   Operational Example 2.   Contrary to the large scale Grid application discussed above, here we consider a very small, but computationally demanding application. The application is CFD simulation of a moving car and the end-user is interested only on the final results. The job request is just to run the simulation on a single machine with a supplied set of data. Further, assume that the service is provided free of charge. Functionally, following sequence of operations will take place:

1 The generic Grid platform announces the service availability (through UDDI).

2 An interested client forwards the job description request to the platform.

3 The SM/CME part of the platform will analyses the job description. The job description will see that that there are no proprietary features and all features to be available inside the platform. The request is positively acknowledged.

4 Having accepted the job, the SM/CME should authenticate the user and authorise the job for further manipulation. SM/CME loads the "AAA" service is available and forwards the request and obtains the response before proceeding further.

5 The policy component need to be loaded and consulted to determine the operational policy, if there are any, to be applied.

6 Once the job is authenticated and authorised, the SM/CME builds a list of resources and features and builds the interaction and dependency map between components or services. In this case, the resource requirements are rather minimal.

7 This in turn requires a consultation with the resource reservation. This necessitates the launching of resource management service which along with the discovery, brokers the requests.

8 In case of reservation is favoured, the SM/CME updates the job status and continues further with the other operations that does not break the dependency or it waits until it is indicated that the resources are available.

9 If the advance reservation was done, then the SM/CME also delegates the provisioning and management tasks to the resource management component.

10 The SM/CME is notified about the resource availability.

11 The SM/CME deploys the data and code on the target machine, by invoking the data management engine.

12 The actual task servicing begins.

13 The SM/CME is notified when the execution finishes.

14 Execution Terminates, modules are unloaded one by one, and the result is forwarded to the client.

4.1.3   Operation Diagram.   Figure 2 shows a very generalized operation diagram of the platform. All Grid services start with the service announcement, which is step 1 in our operational examples. Following the service announcement, the platform may receive number of job submissions along with the job description. This is step 2 in our operational examples. These job descriptions are analysed, authenticated in conjunction with the policy database and accounts database, whichever applicable. Steps 3, 4, and 5 in our operational examples correspond to this. If this operation fails for a reason or another (such as authentication failure), the job submission system (or user) is notified of the failure. If it succeeds, the platform compiles the list of resources, reserves/allocates them for the considered job submission. This requires consultation with the resource reservation system and with the resource discover system. Some of these resources may be secured from remote sites. The step also constructions an action and interaction plan, preparing the task for execution. This corresponds to steps 6-16 in the the first example and 6-11 in the second example. Any failure in any of these operation will result in termination of job processing

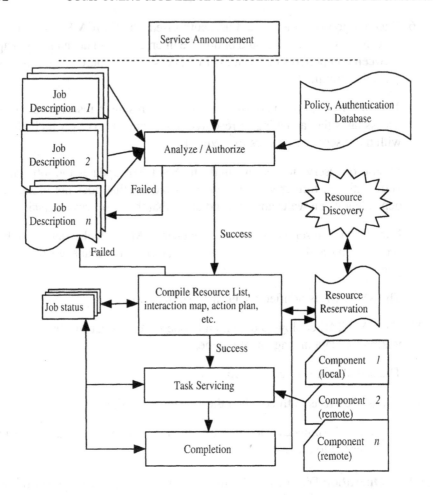

Figure 2.   Generalized operation diagram of the generic Grid platform.

and status of the job is updated. If everything goes well, the actual task servicing begins and this involves component or service invocations and the task will run to termination and status of the job is updated. Thus, the platform dynamically demands and manipulates components as required by the description of each submitted job.

4.2   Advanced Features

From the above examples, it is clear that constantly loading and retaining all the services and features in the platform is not optimal. This is especially true with the case of offering the Grid services in a "plug-and-go" fashion, where there will be large number of small tasks or less demanding applications.

The design and engineering of the generic Grid platform should encompass adaptivity and intelligence. We have not stated this very clearly until this point. The reason for such a delayed introduction of engineering philosophy is to make the cases and requirements very clear.

The engineering of this platform can no longer track the traditional techniques, where the system behaves in a predicted pattern. Instead, the platform should be intelligent and should adapt itself to the changing conditions. These are:

1 As with the case of where applications or customers using their own services (or nominates their service providers), the SM/CME should be able to delegate the operation with those foreign components. This introduces, tremendous amount of freedom and flexibility in the Grid environment. Keeping aside the issue of trust and security, the main challenging issue is the delegation phase. The components inside the Grid system, should adapt themselves (at least to certain extent) to match the interaction with those foreign components.

2 The current state of the Grid system should be taken into consideration by these components. For instance, a pricing component should change the pricing strategy it is using when the system cannot meet the estimated target timing. Similarly, other components need to adapt and re-wire themselves according to the current state.

   Another example is, one component is forced to adapt itself due to a change or adaptation occurring in another component.

3 Next generation application software will be adaptive in nature. This entails that the platform it is relying on also to be adaptive.

4.3   Relevant and Necessary Software Technologies

Section 4.1 has demonstrated the operational principles of the generic Grid platform and the fact that fixed functionality infrastructure would render the Grid platform to be unsuitable for future generation. Most of the software technologies required for engineering the generic Grid platform are readily available for us. This includes Jini [18] for resource discovery, WSDL [6] for service descriptions and so on. The most demanding aspect is the way in which the components have to be built. As illustrated above, the platform is adaptive and requires the components to be adaptive too for effective operations. With this requirement, the interface of service components can vary during the runtime and the SME should be able to understand the interface in order to initiate the interaction. This would not be possible, unless either the SME adapts the interface presented in the *Feature Knowledge Set* to the new set of interfaces announced, or the component itself changes the interface as dictated by the SME.

In a flexible Grid environment, services should be adaptive and the engine should be able to adapt to the current situation, without compromising the security. Such a requirement, where the component interfaces change while they evolve presents us a challenge of interface morphing and/or with the need for dynamically reconfiguring software components [3]. This is well beyond what is offered by Jini [18] or the multi-paradigm frameworks discussed in [12]. However, the techniques for adaptive and customisable software components discussed in [15, 1, 2] can also be applied here. Those approaches tend to provide the necessary level of adaptability required by emerging applications and components such as the one we have outlined above.

5.   Conclusions and Directions for Further Research

The design philosophy described in this chapter still needs to be checked carefully against the available software technologies in order to continue with the next phase – a much more detailed consideration aiming at the development and implementation of a first generic Grid platform prototype. A number of interesting and important issues which need to be addressed in preparation for this next phase include:

- The list of features made available in the feature knowledge set needs to be optimised.

- The proposed design may or may not render the existing protocols and software technologies obsolete. For example, the adaptive nature of the platform requires much smarter discovery protocols. We need to investigate the possible side effects that this feature can have on other components and on the overall design and operation of the platform.

- The adaptive nature of components demands more on the software technology side. For instance, inter-component dependency analysis is a critical part of the adaptive concept. Cross component interaction opens up lots of questions for research, such as cross-component optimisation, security and trust issues to be enforced by the Grid platform.

- The choice of customers or applications providing their own components involve additional challenges. In essence, the platform is provided with a component, which it has never seen. As mentioned in Section 4, this requires dynamic reconfiguration and interface morphing of the components. Related works mentioned in Section 4.3 are not yet mature enough, or at least their capability in addressing this problem is yet unknown.

- The generic Grid services platform provides extra flexibility to the applications or to the end users by permitting them to redefine the feature knowledge set and to update it on the fly. This requires the feature knowledge part to be retained in a separate database, for example an XML file.

- The service manager is a critical component of the proposed platform. We have not looked into the details of the design aspects of this component, and in particular how it directs the CME.

- Techniques for announcing and invalidating interfaces and feature knowledge sets need to be determined.

In addition to the issues mentioned above, there are clearly a lot more that need to be addressed when designing and implementing such a platform in our future work. At this initial stage, we have outlined a possible direction towards building a generic Grid platform using a component-oriented approach. We have addressed in particular the flexibility, longevity, and expandability issues by:

- identifying a core set of features which should build the permanent part of our lightweight Grid services platform;

- identifying an essential feature knowledge set, about which the platform should be aware of;

- separately plugging these components on demand in order to provide an effective operation;

- and by using the reconfigurable software technology to permit these components to adapt themselves to the changing environment needs.

Our motivation is to continue this project and contribute to the development of future generic Grid services infrastructures and corresponding standards.

Acknowledgments

We would like to express our special thanks to Paul Kelly at Imperial College, London for his valuable comments on this chapter.

References

[1] G.A. Agha. Introduction: Adaptive middleware. *Communications of the ACM*, 45(6):30–32, 2002.

[2] M. Astley, D.C. Sturman and G.A. Agha. Customizable Middleware for Modular Distributed Software: Simplifying the Development and Maintenance of Complex Distributed Software. *Communications of the ACM*, 44(5):99–107, 2001.

[3] S. Bagchi, K. Whisnani, Z. Kalbarczyk and R.K. Iyer. The Chameleon Infrastructure for Adaptive, Software Implemented Fault Tolerance. In *Seventeenth IEEE Symposium on Reliable Distributed Systems, (SRDS '98)*, pages 261–270, 1998.

[4] F. Berman, R. Wolski, H. Casanova, W. Cirne, H. Dail, M. Faerman, S. Figueira, J. Hayes, G. Obertelli, J. Schopf, G. Shao, S. Smallen, N. Spring, A. Su, and D. Zagorodnov. Adaptive Computing on the Grid Using AppLeS. *IEEE Transactions on Parallel and Distributed Systems*, 14(4):369–382, 2003.

[5] A. Bosworth, et al. *WS-Addressing Specification*. World Wide Web Consortium, 2003. ftp://www6.software.ibm.com/software/developer/library/ws-add200403.pdf.

[6] R. Chinnici, M. Gudgin, J.J. Moreau and S. Weerawarana. *Web Services Description Language (WSDL) 1.2*. World Wide Web Consortium, 2003. ftp://www.w3.org/TR/wsdl12/.

[7] F. Curbera, et al. Unraveling the Web Services Web: An Introduction to SOAP, WSDL, and UDDI. *IEEE Distributed Systems Online*, 3(4), 2002.

[8] K. Czajkowski, D.F. Ferguson, I. Foster, J. Frey, S. Graham, T. Maguire, D. Snelling and S. Tuecke. *From Open Grid Services Infrastructure to WS-Resource Framework: Refactoring & Evolution*, Version 1.1, March, 2004. http://www-106.ibm.com/developerworks/library/ws-resource/ogsi_to_wsrf_1.0.pdf

[9] K. Czajkowski, D.F. Ferguson, I. Foster, J. Frey, S. Graham, I. Sedukhin, D. Snelling, S. Tuecke and W. Vambenepe. *The WS-Resource Framework*, Version 1.0, March, 2004. http://www-106.ibm.com/developerworks/library/ws-resource/ws-wsrf.pdf

[10] I. Foster, D. Gannon, and H. Kishimoto (Eds). The Open Grid Services Architecture. *GGF-WG Draft on OGSA Spec, Version 19*, 2004. https://forge.gridforum.org/projects/ogsa-wg/

[11] I. Foster, D. Gannon, H. Kishimoto, and J.J. von Reich (Eds). Open Grid Services Architecture Use Cases. *GGF-WG Draft on OGSA Use Cases 2.0*, 2004. https://forge.gridforum.org/projects/ogsa-wg/

[12] V. Getov, G. von Laszewski, M. Philippsen, and I. Foster. Multiparadigm Communications in Java for Grid Computing. *Communications of the ACM*, 44(10):118–125, 2001.

[13] The Globus Toolkit, http://www.globus.org/.

[14] A. Grimshaw, A. Ferrari, G. Lindahl, and K. Holcomb. Metasystems. *Communications of the ACM*, 41(11):46–55, 1998.

[15] F. Kon, F. Costa, G. Blair, and R.H. Campbell. The Case for Reflective Middleware. *Communications of the ACM*, 45(6):33–38, 2002.

[16] T. Lindholm and F. Yellin. *The Java Virtual Machine Specification*. Addison-Wesley, Reading, USA, 1998.

[17] S. Tuecke, K. Czajkowski, I. Foster, J. Frey, S. Graham, C. Kesselman, T. Maquire, T. Sandholm, D. Snelling and P. Vanderbilt (Eds). The Open Grid Services Infrastructure (OGSI). *GWD-R GGF-WG OGSI Spec, Version 1.0*, 2003. https://forge.gridforum.org/projects/ogsi-wg/

[18] J. Waldo and K. Arnold. *The Jini Specification (2nd edition)*. Jini technology series, Addison-Wesley, Reading, USA, 2001.

A SOLUTION FOR ADAPTING LEGACY CODE AS WEB SERVICES

Bartosz Bališ, Marian Bubak, and Michal Wegiel
Institute of Computer Science, AGH,
Kraków, Poland
balis@uci.agh.edu.pl
bubak@uci.agh.edu.pl
mwegiel@student.uci.agh.edu.pl

Abstract   This chapter presents a universal architecture for porting legacy code to Web service environments. We provide a detailed analysis of the proposed solution and characterize it in the context of fundamental Grid requirements. The architecture is evaluated on the basis of such criteria like performance, security, scalability, and fault tolerance. Our solution provides support for process migration, checkpointing, and transactional processing. Both concurrent and asynchronous method-invocation patterns are supported. In addition, we describe a framework that was developed to facilitate the use of the proposed architecture. It reduces implementation effort by automatic code generation. Finally, we present performance evaluation results.

Keywords: legacy software, Web services, Grid services, adaptation, migration, framework

1.   Introduction

This chapter presents a universal architecture which can be employed when adapting legacy applications to Web service environments. The term "Web service", which is extensively used in our discussion, should be interpreted either as OGSI-compliant [12] Grid service or as WSRF-compliant [13] Web service, associated with stateful resources. Both service types are equivalent in terms of the offered capabilities and there is a straightforward mapping between the concepts on which they are based.

We provide a thorough analysis of the proposed solution and assess its characteristics from different perspectives. In particular, the architecture is confronted with fundamental Grid requirements [5], i.e. performance, security, scalability, and fault-tolerance. We also describe a framework which was developed in order to facilitate the implementation of our solution. It comprises a set of tools which allow for automatic code generation. Finally, we provide a performance evaluation of the presented architecture.

This chapter presents the most recent stage of evolution of our solution. Earlier versions thereof are described in [1–2].

2.   Related Work

The issue of adaptation of legacy software to Web service platforms is gradually gaining interest both in scientific and commercial settings. However, presently no comprehensive solutions addressing this area are available. Existing approaches possess numerous limitations and offer poor versatility.

In [9] a proposal of a semi-automatic technique for conversion of legacy C interfaces to their Java equivalents is presented. Two auxiliary tools: JACAW (JAva-C Automatic Wrapper) and MEDLI (MEdiation of Data and Legacy code Interface) are introduced which allow for code wrapping and data mapping, respectively. They employ the Java Native Interface and therefore are restricted to configurations in which legacy applications are located on the same machine as the service container. This solution is also unsafe due to the fact that legacy code is executed within the same operating system process as the container's virtual machine. For example, errors present in a legacy library can manifest themselves by crashing the whole runtime environment.

In [10] a conceptual architecture for adaptation of legacy applications to Web service environments is presented. Three components constitute the essence of the proposed solution: Web service containers, Web service adaptors and back-end legacy servers. Each Web service is equipped with an adaptor which is responsible for connecting to the appropriate backend server on behalf of the clients. The role of adaptors is to hide the complexity of calling backend functions which typically involves communication through proprietary protocols. The most important disadvantage of this approach is its inherent insecurity.

Each backendfor example server demands an open port on which it can listen to the client requests. In complex installations this may introduce serious security vulnerabilities. Another drawback connected with this architecture is its inflexibility. Service adaptors have to be configured statically with regard to the locations of the corresponding backend servers so that the communication can be established. In consequence the infrastructure cannot tolerate process migration between computing nodes and thereby lacks such features like automatic load-balancing and fail-over.

Recently, in [6]an approach to wrapping legacy applications as components based on using the factory pattern was presented. The user provides a script that can execute the application and an XML file that describes the application and the input parameters. It also addresses the security issue.

Our approach allows to overcome the limitations of the above-described solutions.

3. Architecture

We propose a three-tier, client-server architecture in which three main components can be distinguished: the *Service Requestor* (further also referred to as the *Client*), the *Runtime Environment* and the *Legacy System*. They are potentially hosted on different machines. Communication between Service Requestor and Legacy System is mediated by services deployed within the Runtime Environment. Fig. 1 depicts the configuration for a single legacy application exposed as a Web service. It shows the relationships between individual entities along with their cardinalities.

3.1 Service Requestor

Cooperation with Legacy Systems is fully transparent. From the client's perspective, only two Web services are interesting, namely *Factory* and *Instance*, and the other ones are not accessible. Service requestors are expected to follow a specific interaction pattern. Each client shall create its own *Instance* before any operations are invoked. This task is realized with the help of a *Factory*. When processing is finished, no longer needed Instances will be destroyed.

3.2 Legacy System

The legacy system constitutes an environment in which the legacy software is executed. This component plays a crucial role in our architecture since it is responsible for actual request processing. In order to enhance performance and scalability and to improve reliability and fault-tolerance we may install several redundant copies of a single legacy application on different computing nodes, possibly in various geographic locations. For this reason, we can end up with multiple legacy systems associated with a particular Web service.

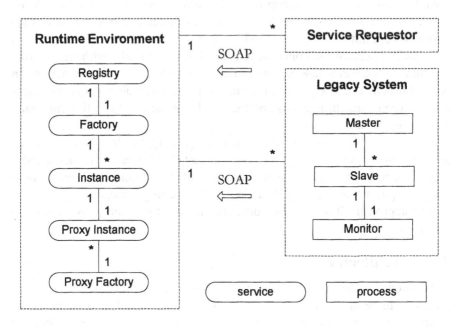

Figure 1.   Proposed architecture

They collectively can be thought of as a dynamic pool of available processing resources.

The central concept that we propose is that legacy systems do not operate as network servers, but they are designed to behave as Web service clients. They receive requests and deliver responses by calling dedicated Web service operations. This considerably enhances security and makes process migration feasible.

In the context of a particular Web service, a legacy system comprises three types of processes which are: *Master*, *Slave* and *Monitor*.

3.2.1   Master.   The master process is a one-per-host, permanent entity. Its primary responsibility is the legacy system registration. Whenever the load of the machine on which the master process is executed allows to serve a new client, the master process reports this fact by calling the registry service. Along with this invocation, the estimated processing capabilities and validity timestamp are provided. The call blocks the master process until the specified time limit expires or one of the clients is assigned. In the latter case, the master process spawns new slave and monitor processes which take over further serving of the newly assigned client.

3.2.2   Slave.   The slave process is a transient entity, always associated with a certain client. Its lifetime is limited to the period during which the client interacts with the Web service. The number of concurrently running slave processes is changing as clients come and go. Slave processes are in charge of direct cooperation with the legacy software. This can range from invoking functions from local system libraries to network communication over proprietary protocols. Slave processes are responsible for retrieving client calls, translating them to the legacy interface, performing actual processing, and delivering the obtained results. This is achieved by means of blocking invocations of methods belonging to the proxy instance interface.

3.2.3   Monitor.   Each monitor process is associated with a certain slave process (one-to-one relationship). The monitor is responsible for generating a heartbeat signal and it repetitively calls a special method on the corresponding proxy instance. This allows to:

- assure both sides that the network connection is working,

- inform the proxy instance about the current status of the slave process,

- cancel the execution of the slave process at any time.

To show the functionality of the monitor process let us consider the most typical failure scenarios:

- connection is broken – in this case the monitor terminates the slave process which otherwise would continue to use system resources until request processing is finished,

- slave process crashes – in this case the monitor informs the proxy instance about this fact which otherwise would have to wait until request processing times out.

The justification for making the monitor a separate process is as follows:

- Slave processes execute legacy code which can contain errors leading to abnormal process termination or even non-termination. The detection of such situations has to be delegated to another process.

- In order to make the design elegant and the implementation cleaner we separate processes which fulfill different roles. Moreover, merging master and monitor processes would result in a multi-threaded process which could unnecessarily complicate the implementation.

3.2.4   Middleware.   An important issue is how our architecture relates to the job submission mechanism and resource brokering facilities developed along with the Grid infrastructure. We employ job submission in order to manage the pool of available master processes. Whenever the number of concurrently connected clients causes that processing capabilities of currently registered master processes become insufficient, a new master process is created. This is accomplished via job submission. Similarly, when there is a significant decrease in the load, one or more chosen master processes are terminated. Thus we fully employ the available Grid middleware. In fact we build on top of the job submission facility in order to provide additional, higher-level functionality. Basic job submission mechanisms allow to execute the specified program for the given input data (batch mode). This is not exactly what we need since we require a conversational interaction with the process that is executed and which potentially maintains internal state.

The main reason for the existence of master processes is enhanced performance. Job submission mechanisms are much slower than spawning a new slave or monitor process. Submitting each separate process via a resource broker would be suboptimal in terms of overhead. Another argument supporting master processes is the possibility to bypass job submission altogether. In small intra-Grid environments, it may prove sufficient to maintain a static configuration in which selected hosts run master processes permanently (for example, they can be started by system scripts when a machine is booted).

One remaining point is why we do not employ notifications instead of blocking interaction with registry and proxy instances. Processes executed within a legacy system could subscribe to notifications instead of repetitive blocking on synchronous invocations. However, the problem related to this approach is that, according to its specification, a notification sink is required to expose a network accessible endpoint. In consequence, legacy systems would have to allow incoming connections in their ephemeral port range and in fact act as servers. This can prove problematic in case of pre-configured firewalls. Moreover, open ports pose a vulnerability that cannot be accepted when security-sensitive applications are used. Another argument against notifications is that they effectively disable process migration. Processes which listen on a specific network socket cannot be transparently moved to a machine with a different network identity. Our approach is free of both these limitations.

3.3   Runtime Environment

The runtime environment maintains a collection of Web services that encapsulate the interaction details with the legacy systems. For each legacy application, there are three permanent services deployed: a *registry*, a *factory* and a *proxy factory*. Depending on the number of simultaneously served clients the number of transient services varies, namely *instance* and *proxy instance* services, which are in one-to-one relationship. Transient services are instantiated by the corresponding factories and are owned by their creators.

Access to all services is granted on the basis of authentication and authorization procedures. Only entities holding adequate identities can invoke a particular operation. In case of internally used services, namely registry, proxy factory and proxy instance, host certificates are employed. For the remaining services, namely factory and instance, user certificates are engaged. Service requestors can access only those instances that they own.

The registry Web service is responsible for controlling the one-to-one mapping between service requestors and legacy systems. It provides the interface that can be employed to:

- assign one of the registered legacy systems to a pending client,

- advertise that the legacy system is ready to process requests.

The registry maintains a priority queue in which volunteering legacy systems are remembered and sorted according to their processing capabilities. This criterion decides which legacy systems and in which order will be assigned to the consecutively appearing clients. When in a particular point in time no master processes are available, an error is returned to a client that tries to create a new instance.

Apart from the registry, we distinguish between two types of services: ordinary and proxy ones. The former are used by clients whereas the latter are designed for internal purposes. The aim of this separation is to achieve a higher degree of transparency. Ordinary services contain only those methods that are interesting to the clients. The proxy instance is the main contact point with legacy systems. Its methods are called by slave and monitor processes. The instance service is forwarding client requests to the associated proxy instance.

4. Scenarios

There are several independent scenarios that can be handled by our architecture under various circumstances. They can be divided in two classes, depending on whether they originate from the client side or the legacy system side.

4.1 Client Side

Fig. 2 presents a diagram that schematically illustrates client side scenarios. All of them are triggered by client requests.

4.1.1 Instance Construction. The instance construction scenario involves two major steps: creation of the associated proxy instance and assignment of one of the registered master processes. It is executed in response to a

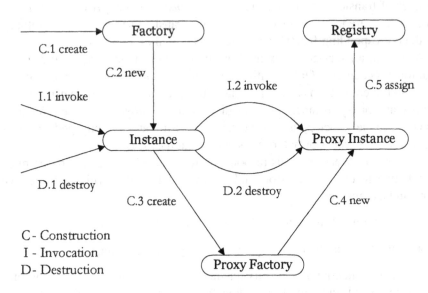

Figure 2.   Client side scenarios

create request sent to the factory service. This scenario is in principle based on the mechanism of lifetime management callbacks which enables to execute custom actions upon service construction. As shown in the diagram, three permanent Web services, namely factory, proxy factory and registry, are engaged in the whole procedure. Upon successful completion, two transient Web services are created: instance and proxy instance. They can be treated as an exclusive property of a particular client. In addition, one of the available legacy systems is assigned. It will intercept client requests, and deliver the corresponding responses.

4.1.2   **Operation Invocation.**   Whenever a client invokes a particular method, full description thereof together with the passed parameters is forwarded to the proxy instance. If synchronous invocation mode is used, the client is blocked and waits until the legacy system delivers the response. In case of asynchronous mode, the call returns immediately and the results are sent later by means of a notification message. Efficient method invocation is crucial to the system performance because it is the most frequently occurring event. Communication between the instance and its proxy instance takes place within a single runtime environment so it is unlikely to pose a bottleneck. The main source of overhead is the cooperation with the legacy systems.

4.1.3   **Instance Destruction.**   The Destruction scenario can be triggered either by an explicit client request or by the runtime environment when the instance's time to live expires. In both cases we rely on lifetime management callbacks. They are employed to send destruction request to the associated proxy instance so that both services are deleted simultaneously.

4.2 Legacy System Side

Fig. 3 presents an interaction diagram for the legacy system side scenarios. For simplicity, we assume that there is already exactly one master process registered (details of its creation, manually or via a job submission mechanism were omitted). The legacy system side scenarios take place automatically and are beyond the control of Web service clients.

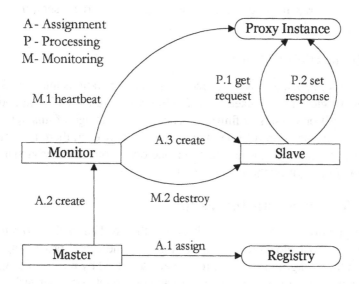

Figure 3. Legacy system side scenarios

4.2.1 Client Assignment. As we discussed earlier, a master process repetitively offers its participation in processing by calling the registry service. When a particular offer is accepted, the invocation returns the endpoint address of the assigned proxy instance. Following this, the master process spawns a monitor process which in turn creates a new slave process. Next, the master process checks whether it can serve one more client, accordingly adjusts the estimation of its processing capabilities and either continues volunteering or waits until the machine load decreases.

4.2.2 Request Processing. The processing of client requests is implemented by two alternately repeated invocations: one collecting subsequent requests and one delivering the corresponding responses. Between them, legacy processing takes place. The slave process is solely responsible for this scenario. In case of any communication error, the slave terminates itself. This does not, however, take place immediately, but only when a long-lasting connection loss is detected.

4.2.3   System Monitoring.   The monitor process periodically calls the proxy instance to report on the current status of the associated slave process. When a connection error is detected or processing is cancelled (for instance because of migration), a termination signal is sent to the slave process. When the slave process exits prematurely for any other reason, the monitor process tries to report this fact to the proxy instance and, having done that, terminates itself. This allows for automatic re-claiming of resources, should any error occur.

5.   Invocation Patterns

Most commonly, the legacy interface comprises operations intended for s ynchronous, single-threaded execution. Such methods run in isolation and block the caller until processing is finished. The main advantage of this approach is that no concurrency-control scheme is required. However, there are situations in which it is necessary or desired to invoke operations asynchronously, or to execute a number of methods in parallel.

5.1   Asynchronous Invocation

The asynchronous execution mode allows the invoked method to return immediately without blocking the calling thread. The actual processing is performed in the background and the results are delivered by means of a signaling mechanism. In the meantime the caller can concentrate on other tasks. In our framework, asynchronous method invocation is supported and implemented on the top of notification facility. Requestors can interact with the selected methods according to the notification source/sink design pattern. The only change in the generated code for enabling this feature is necessary for the instance and the proxy instance services. The protocol for cooperation with the legacy systems is not affected since it always takes place in asynchronous manner.

5.2   Concurrency

Occasionally, service requestors may need to execute several operation invocations at the same time. This may be caused by two different situations:

- several synchronous calls are made by concurrently running threads,
- several asynchronous calls are made in short succession.

Both cases introduce concurrency to the legacy code wrapper. In order to provide support for simultaneous execution of a number of methods, the slave processes need to start a new thread for processing each client request. In consequence, synchronization mechanisms have to be applied to avoid race conditions caused by concurrent access to the shared data. Our framework supports the development of thread-safe code, nonetheless programmers providing the mapping to a legacy interface that allows concurrent invocations are strongly recommended to give serious thought to their implementation. Since parallel method execution considerably complicates development, it is generally a good design practice to avoid it whenever possible.

6. Process Migration

The main motivation behind automatic process migration is enhancing fault tolerance and load balancing. These two aspects are of paramount importance to the Grid environment. Process migration is indispensable when we need to:

- dynamically offload work onto idle machines,

- transparently recover from system failures.

6.1 Migration Techniques

There are two broad classes of process migration solutions: low-level and high-level ones. The low-level approach is limited to machines with the same architecture and can be employed only in homogenous environments (e.g. clusters of workstations). This is because such process migration relies on the transfer of binary state which includes, but is not limited to, a virtual memory image and current contents of processor registers. Low-level migration is fully supported in the proposed architecture as slave processes act as clients rather than as servers. They can be moved to another machine at any time since their network identity is not required to remain unchanged. Legacy applications are free to take full advantage of migration capabilities present in the operating system under control of which they are executed.

The high-level approach is in principle based on the replay of the sequence of operations, constituting method invocation history. This can be an expensive way of state restoration but it is applicable in heterogeneous environments. In order to be able to re-iterate through the past operation calls we need to store all information regarding method invocations as they come from requestors. It seems that we should also remember the output of each executed method so that at a later time we could compare it with the results returned by calls issued during state reconstruction. This is essential since we must ensure that the repeated scenario proceeds exactly like before, because when migrating we are acting on the requestors' behalf without their knowledge.

One of the problems with history-based state restoration is the high memory consumption. This can be tackled by using disk storage. However, dumping history into a file is a very slow operation. Therefore we should employ buffering as much as possible.

6.2   Point of Migration

Most frequently, a migration procedure is triggered when some erroneous state is entered. We conceptually distinguish between two types of failures:

- logical errors – caused by invalid parameters supplied by the user,

- system errors – caused by network malfunction or hardware crash.

An attempt to recover from a logical error by means of process migration does not make much sense. In such circumstances, the failure cannot be masked and its description has to be delivered to the requestor. In the proposed architecture, a system error is recognized when the number of lost keep-alive messages exceeds a specified limit. This phenomenon indicates that the currently assigned legacy system cannot be contacted and another one should take over request processing. Therefore, the registry service is invoked and upon successful re-assignment, state restoration takes place.

Process migration might also prove to be a good solution when a legacy system's response time unexpectedly increases drastically. Such a situation can be caused by a temporal peak in resource utilization. Delegating work to another machine may help to evenly disperse the load. Putting this mechanism into operation requires that a maximal execution time can be estimated for each method. Moreover, it is necessary to interrupt and cancel the processing of the currently assigned slave process. In our architecture, this can be achieved using the monitor process which periodically invokes a heartbeat method. This call returns a boolean value which, when set to false, tells the monitor to terminate its subordinate slave process.

6.3   Optimizations

It is rarely necessary to remember the sequence of all operation invocations. For instance, stateless applications can be restarted after a failure without any concern for their history of prior interactions. Most services maintain an internal state; nevertheless, it is usually possible to apply various optimizations concerning the amount of data that needs to be saved. For example, we may know that execution of certain methods resets the state and allows to clear the history. For this purpose we provide a special method in the interface of the proxy instance which should be called whenever a given invocation sequence can be safely discarded.

In our framework, high-level process migration can also be guided by transactions and checkpointing which can work both separately and in combination.

6.4 Checkpointing

Certain legacy applications support checkpointing. That means that they are capable of periodically saving a snapshot of their state. This enables efficient state restoration after process restart. Our framework provides facilities for exploiting checkpointing mechanisms if they are available. A dedicated interface is implemented for this purpose. It is accessible only on the legacy system side and allows:

- saving the current state snapshot,

- retrieving the most recently saved state.

We assume that legacy applications are able to represent their state in the form of a string. State snapshot is stored on the container side (possibly on disk) and its contents are not interpreted in any way. It is a recommended practice to check whether any valid state dump is available when an application starts. This allows to eliminate the need for slow state reconstruction based on invocation history. For this solution to work properly, applications are required to provide state snapshots regularly, whenever they undergo a major change in state, especially if they are computationally expensive.

It is noteworthy that checkpointing itself does not eliminate the need for history-based state restoration. This is because system failure may occur between subsequent checkpoints, when some invocations following last checkpoint are already made. In such a situation, in order to keep the migration transparent, it is necessary to load the latest state snapshot and then to repeat the remaining short sequence of method calls. For this reason, checkpointing cannot work properly in isolation and needs to be combined with the technique of method invocation replay.

6.5 Transactions

Legacy systems that operate in a transactional fashion require special assistance from our framework. This is particularly important in the context of process migration. A sequence of operations constituting a single transaction is by definition atomic. In the event of transaction failure, none of its components should have any effect. This implies that aborted transactions do not affect system state and, in consequence, should be discarded during state restoration that takes place upon migration.

Our framework provides operations for starting, committing and aborting transactions. The transaction management methods are present in the interface of instance. A collection of operations that form a single logical unit of work should be surrounded by appropriate invocations. When no explicit transaction is started, it is assumed that each operation constitutes a separate transaction which is auto-committed upon successful completion. If method invocation fails for some reason, the currently executed transaction is marked as aborted. Only committed transactions are recorded. Aborted ones are excluded from invocation history since they introduce no change in state.

7.   General Properties

The proposed architecture satisfies the requirements that ought to be met in Grid environments. It possesses many desirable features as discussed below.

7.1   Security

There are two aspects concerning the security of our solution:

- it is not necessary to introduce open incoming ports on the machines where legacy software resides,

- it is possible to authenticate the machines with which we cooperate and verify that the processing is delegated only to trusted nodes.

Both these advantages are due to the fact that processes executed within legacy systems act as clients rather than servers. We rely on the security infrastructure provided by the given runtime environment. Thus, depending on the particular application, various security mechanisms can be used. By default, we perform authentication and authorization procedures. If needed, communication integrity and privacy can be ensured by means of digital signatures and encryption, respectively.

In our architecture, the security configuration can be thought of as two lists of identities:

- for clients that are entitled to use our service, and,

- for hosts that are permitted to register in the context of our service.

In consequence, maintenance of security policies should not involve much administrative effort.

7.2   Scalability

The combination of several factors contributes to good scalability of our architecture:

- Processing is highly distributed since all tasks are delegated to legacy systems.

- Services deployed within the runtime environment do not consume much resources as their activity is restricted to message forwarding.

- Job submission mechanisms are employed which enable dynamic resource allocation in response to unexpected changes in utilization.

- Automatic load balancing is ensured by master processes advertising processing capabilities of the machines on which they are executed.

7.3 Fault Tolerance

The proposed architecture offers a high degree of immunity to component failures for the following reasons:

- Both low-level and high-level process migration are supported.

- Monitor processes generate heartbeat signals, which enables fast detection of failures.

- Support for checkpointing enables fast state restoration.

- Support for transactions allows recovery from uncommitted operations.

- A registration model is used that enhances responsiveness to sudden changes of configuration.

7.4 Versatility

An important advantage of our architecture is the fact that we make no assumptions regarding programming language or platform on which its individual components are based. Our solution is universal enough to accommodate a variety of legacy systems and runtime environment implementations. Furthermore, legacy software can remain in the same place where it was initially installed. There is no necessity of moving programs between machines or changing their configuration. No modifications of legacy code are required (non-intrusiveness). The net effect is that our architecture can be applied in a wide range of different adaptation scenarios.

8. Implementation

We have developed a framework comprising a collection of Java tools facilitating the adaptation of legacy C/C++ applications to the proposed architecture. At present, only OGSI-compliant services are supported since we employ Globus Toolkit 3.2 [7]. We plan to migrate to WSRF as soon as its first implementation becomes available. The core functionality provided by our framework can be described by its typical use case, which is presented below.

1   A developer specifies the Java interface that will be exposed by the deployed service. In case of complex data structures this may also involve definition of accompanying Java classes. Usually, the provided interface mirrors or at least resembles its legacy equivalent.

2   Source code generation takes place. This includes the creation of Java and C++ classes as well as the required deployment descriptors and build scripts. Developers can override default settings in order to customize aspects like concurrency or transactional mode.

3   The developer provides the implementation for methods comprising the generated C++ interface (which effectively is the translation of the earlier specified Java interface) in order to define the mapping to the legacy interface. This is the only phase which may involve development effort. However, unless legacy and service interfaces differ considerably or concurrent method invocation is enabled, this task should be straightforward.

4   Auto-generated build scripts take care of building a deployable package and compiling C++ sources to executable programs.

In our current implementation we employ gSOAP 2.4 [8]. Operation invocations performed by service requestors are forwarded to the legacy systems in a serialized form. This allows us to uniformly treat all methods regardless of their formal parameters and returned values. A special data format was devised for this purpose.

When designing our framework we have put particular emphasis on two aspects:

- universality, so that a wide range of different legacy applications is supported,

- ease of use, so that developers have to concentrate only on the most important things.

Since these aims are often in conflict, we had to make many tradeoffs. One of the test cases for the framework implementation which is worth mentioning was the adaptation of the OCM-G [4] Grid application monitoring system to Globus Toolkit 3.0. The OCM-G works in an event-driven manner, therefore an asynchronous, concurrent programming model had to be employed.

A prototype version of our framework (having about 4000 lines of code in Java and C++) offers support for most features of our architecture (including process migration, transactions and checkpointing). The remaining functionality is successively added. Currently, our work is primarily focusing on three aspects:

- development of additional test cases,

- refactoring of current implementation,

- providing more complete support for our architecture.

9. Performance Evaluation

In order to estimate the communication overhead introduced by our architecture, we conducted an experiment which aimed at comparing the performance of two Web services, one of which was dependent on a legacy system. They both offered the same functionality as seen from the client perspective. Specifically, each service was exposing a single operation which was returning the length of the string passed on as its parameter. Construction and destruction scenarios were excluded from the measurement because they are always executed once as opposed to potential multiple method invocations.

We used the metrics of bandwidth and latency. Transmission time depends linearly on the message length and is given by the following formula:

$$time = \frac{length}{bandwidth} + latency$$

where the two above-mentioned quantities are constants.

The experiments were carried out on a single-processor IA-32 machine running the Linux operating system which simultaneously hosted all three components of our architecture. No security mechanism was employed, i.e. neither authentication nor authorization was performed. In consequence, the obtained results reflect the overhead introduced solely by our architecture. The influence of different security mechanisms on efficiency of Globus I/O is characterized in [3]. All tests were performed on the client side.

We measured the time needed to execute a single operation call for data payload ranging from 0kB to 50kB (with the granularity of 1kB). For each message length, we calculated an average method execution time on the basis of a series consisting of 100 consecutive invocations (Gaussian distribution was assumed). The obtained measurement results are presented in Fig. 4. There, the *ordinary service* is the one which is independent of legacy systems.

The calculated values of parameters are listed in Tab. 1. We can expect around 2.5-fold increase in transmission time when a Web service is backed up

Table 1. Calculated values of bandwidth and latency

Quantity	Ordinary service	Legacy service	Relative change
Bandwidth	909.1 kB/s	370.4 kB/s	reduced 2.5 times
Latency	15.4 ms	37.8 ms	increased 2.5 times

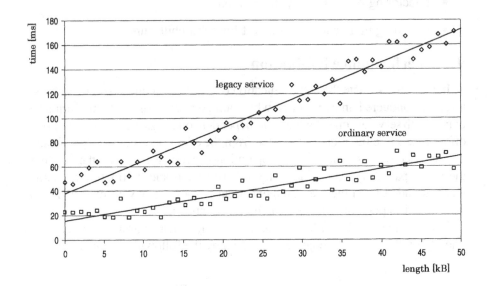

Figure 4.   Measurement results for method invocation scenario

by a legacy system. This stems from the fact that in case of a legacy service each method call needs to be first forwarded to the proxy instance and then to the legacy system. Such overhead is, however, perfectly tolerable for applications in which communication does not dominate computations.

10.   Conclusion and Future Work

In this chapter we have discussed the techniques of migration from legacy software to Web service environments. A proposal for a versatile solution compliant with Grid requirements was presented.

We have shown how to implement our architecture on top of available Grid middleware. Moreover, we presented our first experiences with a framework, the prototype of which has been developed. We proved that despite the relative complexity of the proposed architecture, developers employing our framework can readily adapt their C/C++ software to our Web services platform. We evaluated the communication overhead introduced by our architecture, and demonstrated the obtained experimental results. We drew a conclusion that this overhead is tolerable in case of computationally intensive applications.

We intend to continue the development of the presented architecture and add support for:

- redundant parallel processing (multiple legacy systems are assigned to the same task in order to improve reliability),

- early process migration (migration is started in advance when quality of service begins to decrease so that in the event of failure state restoration can be accomplished faster),

- real-time processing (method invocations are queued according to their priorites).

The implementation of the framework is planned to follow the appearing architectural extensions. More information about this research is available at our project Web page [11].

Acknowledgments. This research is partly funded by the European Commission IST-2001-32 243 Project "CrossGrid" and the Polish Committee for Scientific Research SPUBM 112/E-356/SPB/5.PR UE/DZ224/2002-2004.

References

[1] B. Baliś, M. Bubak, and M. Węgiel: Migration from Legacy Software to Grid Services. Proc. *Third Cracow Grid Workshop*, Cracow, Poland, 2003, pp. 254–266.

[2] B. Baliś, M. Bubak, and M. Węgiel: Adaptation of Legacy Software to Grid Services. Proc. *International Conference on Computational Science* 2004, Cracow, Poland, Springer LNCS 3038, pp. 26–33.

[3] B. Baliś, M. Bubak, W. Rząsa, T. Szepieniec, and R. Wismüller: Two Aspects of Security Solution for Distributed Systems in the Grid on the Example of OCM-G. Proc. *Third Cracow Grid Workshop*, Cracow, Poland, 2003, pp. 197–206.

[4] B. Baliś, M. Bubak, W. Funika, T. Szczepieniec, and R. Wismüller: Monitoring Grid Applications with Grid-enabled OMIS Monitor. Proc. *First European Across Grids Conference*, Santiago de Compostela, Spain, February 2003, Springer LNCS 2970, pp. 230–239.

[5] I. Foster, C. Kesselman, J. Nick, and S. Tuecke: The Physiology of the Grid. An Open Grid Service Architecture for Distributed Systems Integration.
http://www.globus.org/research/papers/ogsa.pdf

[6] D. Gannon, S. Krishnan, A. Slominski, G. Kandaswamy, and L. Fang: Building Applications from a Web Service Based Component Architecture. In: *Component Models and Systems for Grid Applications*, pp. 3–17, Springer, 2004.

[7] Globus Project: http://www.globus.org

[8] gSOAP Project: http://www.cs.fsu.edu/~engelen/soap.html

[9] Y. Huang, I. Taylor, D. Walker, and R. Davies: Wrapping Legacy Codes for Grid-Based Applications. Proc. *International HIPS Workshop*, Nice, France, 2003.
http://www.cs.cf.ac.uk/user/I.J.Taylor/CV/Papers/MedliHIPS2003.pdf

[10] D. Kuebler, and W. Eibach: Adapting Legacy Applications as Web Services, IBM Technical Report.
http://www-106.ibm.com/developerworks/webservices/library/ws-legacy/

[11] LGF – Legacy to Grid adaptation Framework: http://www.icsr.agh.edu.pl/lgf/

[12] Open Grid Services Infrastructure: http://www.gridforum.org/ogsi-wg/

[13] Web Services Resource Framework: http://www-fp.globus.org/wsrf/

II

MIDDLEWARE ARCHITECTURE

A GRAPHICAL MODELING ENVIRONMENT FOR THE GENERATION OF WORKFLOWS FOR THE GLOBUS TOOLKIT

Francisco Hernández, Purushotham Bangalore, Jeff Gray, and Kevin Reilly
Department of Computer and Information Sciences,
University of Alabama at Birmingham,
Birmingham, AL, USA
hernandf@cis.uab.edu
puri@cis.uab.edu
gray@cis.uab.edu
reilly@cis.uab.edu

Abstract   Grid computing aims at managing resources in a heterogeneous distributed environment. The Globus Toolkit provides a set of components that can be used to build Grid-enabled applications. Presently, applications are typically hand-crafted either by using a set of command line interfaces, or by using a set of Java packages provided by the Java CoG Kit. The purpose of this work is to introduce a high-level layer that abstracts and simplifies the development of applications within the Globus Toolkit context by creating graphical workflows of applications using domain-specific modeling techniques.

The expected impact of this effort is a reduction of the development time involved in generating applications for the Globus Toolkit. An additional advantage is to provide a high level view for the construction of Grid applications using the Globus Toolkit that avoids some of the intricacies documented for other approaches. Furthermore, the concepts introduced in this chapter can be employed not only in the context of the Globus Toolkit but with other component frameworks..

Keywords: domain specific modeling, workflows, Globus toolkit, Java CoG kit, code generators, visual authoring tools, automatic programming, software engineering

1. Introduction

The Globus Toolkit [13] is the de facto standard for building Grid-enabled applications. A user can choose three different approaches to construct such applications: (1) utilize a command-line, (2) exploit a C [25] API, or (3) employ a commodity toolkit such as the Java CoG Kit [32]. All of these approaches require an in depth understanding of the underlying technologies involved in constructing Grid applications. This limits the use of the Toolkit to those who are knowledgeable about the intricacies of these technologies. Traditionally, Problem Solving Environments (PSE) or portals [28] have been developed to ease the construction of Grid applications. PSE's provide a high-level view for specifying Grid-enabled applications and rely on middleware to connect with the Grid component resources [14]. This kind of tool expedites simple tasks (e.g., simple job submissions, and checking the status of a previously submitted job), but it lacks the flexibility to define a complex sequence of tasks.

Workflows have gained increasing attention for their application in composing a flow of tasks in a Grid environment [33]. Workflows describe the execution of complex applications built from individual application components, which is similar to the process used to construct applications using the Globus Toolkit. Previous workflow studies vary in complexity, ranging from the use of artificial intelligence to handle the automatic creation of workflows [9, 34], to the specification of grid flows using an XML file [6, 12, 31].

Pegasus [9], and GridAnt [31] deserve special attention because they can be considered the endpoints of workflow approaches. Pegasus uses complex artificial intelligence planning techniques to generate automatically resource mappings and tasks according to application goals. GridAnt, on the other hand uses the Apache ANT tool [5] as a basis for its workflow engine. Although these two tools offer a viable solution to the workflow specification, they are rather difficult to use for a new Grid user. The level of technological complexity in Pegasus and the XML input requirement in GridAnt make it difficult to specify the workflow.

A solution is needed that removes these accidental complexities of use and embeds experimental knowledge of the domain into a code generator that can generate the complex configurations. Such a technology exists in the area of domain-specific modeling [21]. With this technology, a user focuses on higher levels of abstraction at the problem space and is able to avoid low-level details, such as Grid services and their usage.

The approach used in this chapter is based on the concept that the development of Grid enabled applications can be improved by mapping the different Globus components into entities of a graphical model. The graphical models compose a high-level layer that abstracts and simplifies the development of Grid applications by providing all the capabilities of Globus but hiding all the low-level implementation details. The mapping between Globus components and the graphical models is performed by using concepts of domain-specific modeling that utilizes the interfaces provided by the Java CoG Kit. By combining these graphical entities, a particular application workflow can be generated into the Java [19] code that utilizes the Java CoG Kit API. Three research issues are exploited in this chapter:

1 The creation of a meta-model that maps the Globus Toolkit's components to a graphical model. This meta-model defines the language used to construct workflow models.

2 The generation of graphical workflows between the different tasks of the application through the use of the meta-model.

3 The generation of Java programs from the graphical workflows. This is realized by using a model interpreter that traverses the graphical workflows and generates a program that manages the application execution.

The techniques presented in this chapter abstract the component model in an independent way. Such techniques can be applied to different component frameworks. By creating different model interpreters, one for each component framework, the same graphical elements can represent components from different frameworks. These components can then be mixed and code for the corresponding framework can be generated. The rest of the chapter is organized as follows: Section 2 provides background on building applications with the Globus Toolkit and Java CoG Kit; Section 3 introduces the methodology that is used; Section 4 presents related work; Section 5 presents future work to be explored; Section 6 offers conclusions of the present work.

2.   Background

The Globus Toolkit [13] provides a common middleware that considers resources as entities of a virtual organization. This facilitates the construction of Grid applications. The middleware is formed by different components such as the Globus Resource Allocation Management component (GRAM) [8], and Grid Information Services (GIS) [7], which provide services to integrate distributed resources in a Grid computing environment. Creating an application that uses this Toolkit requires the composition of several of these components.The interaction with these components is accomplished by using a simple interface that permits the manipulation of the underlying low level resources. Globus does not enforce any particular programming model so different applications or Grid tools can be constructed using this set of components. Furthermore, an application builder can use only the components that are required for his application and incrementally incorporate additional components to make his application more Grid-aware.

Originally, two approaches were used to generate this composition: (1) a set of command-line tools, and (2) a C API provided by the Toolkit. With these two approaches a user can interact with the interface provided by Globus and manipulate the low level resources. These two approaches provide an ideal solution for an experienced Globus programmer by allowing him to optimize the use of the resources, but for an inexperience user this programming model increases the complexity required to write an application. Considering the dynamic behavior of a Grid system, both approaches are less than satisfactory if the user is not a Globus savvy programmer [34].

One solution to this problem is to create a layer that sits on top of the components provided by Globus. Such a layer is provided by commodity toolkits. The Java Commodity Grid Toolkit (Java CoG Kit) was created to assist in the development of applications using Globus Toolkit [32] services; it was a step towards simplifying the construction of applications for the Globus Toolkit. The Java CoG Kit helps a user navigate the intricacies of the Globus components more easily by introducing a new programming model for the Grid. Furthermore, the Java CoG Kit provides many utility components organized as Java Packages that enhance the functionality of Globus. However, although the Java CoG Kit improves the interface between users and the Globus components, even the user who is Java savvy still needs to dedicate additional time in order to learn how to build applications for the Globus Toolkit. A method that incorporates these widely-used technologies in a more accessible and efficient manner can be achieved using concepts of domain-specific modeling.

In domain-specific modeling, a design engineer creates models for a specific domain using concepts and terminology from that domain [20]. The domain-specific models are developed by first creating a meta-model that specifies the ontology of the domain. The meta-model serves as a paradigm, or language, that defines the syntax and static semantics for models of that domain; the dynamic semantics are introduced by an interpreter that synthesizes the models into different representations [24] (e.g., XML configuration files or source code). The Generic Modeling Environment (GME) [26] is a graphical tool that automates the creation of domain-specific models. GME allows a user to create graphical models by providing a general paradigm (i.e., language) from the meta-model definition.

3.   Methodology to Support Model-Driven Generation of Workflows

Two actions are necessary to create domain-specific models for the Globus Toolkit:

1 Definition of the meta-model, defining the paradigm (language) to be used to create workflow models.

2 Implementation of the interpreter that translates the workflow models into corresponding Java code.

Both of these actions are implemented using GME. One of the advantages of using GME is that it allows a modeler to define base elements that can be reused in more complex models. This property is a major advantage because it is possible to define elements such as resources, user credentials, file transfers and job submission tasks only once and then reuse them in any specification of a workflow.

The models created can then be translated into executable specifications used to synthesize automatically various software artifacts [29]. The translation is performed by a model interpreter that recognizes the concepts from the workflow language and generates the semantic actions associated with that concept. In the following subsections, the manner in which the meta-model and its interpreter are constructed is explained.

3.1   Meta-Model

The goal of the meta-model is to define a new visual language that can be used to create specific workflow models. The design of the meta-model is based on the experimental knowledge of the particular domain. In this case, the design is based on the manner in which a user specifies the sequence of tasks in an application's workflow.

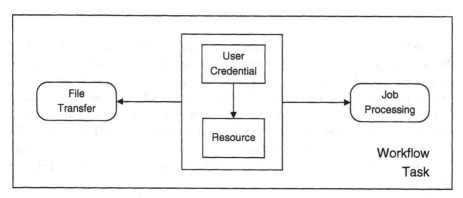

Figure 1.   Workflow task specification. A workflow task consists of file transfers or job processing tasks. These two tasks use a specific Grid resource. Resources require an authentication mechanism provided by the user's credentials.

Figure 1 illustrates the manner in which workflow tasks (subsequently referred to as "tasks") are defined. The central rectangle indicates that each resource requires an authentication method given by the user's credentials.

Rounded rectangles specify tasks consisting of file transfers and job processing. According to the requirements specified in Figure 1, four aspects need to be considered in defining the meta-model:

1 Resources for running jobs and performing file transfers, including the specification of the credentials required to authenticate the resources.

2 File transfers end-points, including resource, location on resource, and file name.

3 Jobs, including their resource and input parameters.

4 Workflows, which are a composition of the previously defined tasks.

The definition of these four aspects provides a *mapping among* the basic requirements for constructing grid applications, the services provided by Globus, and GME entities. The way in which these aspects are specified in the meta-model and their corresponding use is explained in the following subsection. Additional details can be found in [22].

3.1.1   Meta-Model Construction and Example Usage.

For explanation purposes, the meta-model can be subdivided into four different parts (Figures 2 and 3), corresponding to the aspects enumerated in the previous section. All parts are specified using the same entities provided by GME's general paradigm. The distinction of the GME entities in each part is demarcated by the concept that each entity represents in the domain, and the relationship between these concepts.

The basic concepts in the domain are Resources, File Transfers, Jobs, and Workflows. These are defined in the meta-model using either an (GME) Atom or a (GME) Model entity. Model entities can contain other model entities or atoms, but atoms are indivisible. File transfers and jobs are defined as Model entities because they contain resources. Both of these concepts require state information, so (GME) attributes are associated with these entities (Figure 2.b and 3.a). For example, the definition of a job task requires the specification of its RSL (Globus Resource Specification Language) [18] parameters (Figure 3.a).

The association between a resource and its corresponding authentication credentials is given by a (GME) connection entity. An attribute that indicates if the resource is local or remote is associated to the Resource atom. Resources that are remote need an authentication credential. Resources can be used either for computation or for data storage. As typical in UML models, the triangle of Figure 2 indicates that both kinds of resources inherit attributes and connections from a basic host entity (Figure 2.a). Finally, the workflow part of the meta-model consists of the previously defined tasks (file transfers, and job specifications), and a start and end of workflow markers (Figure 3.b).

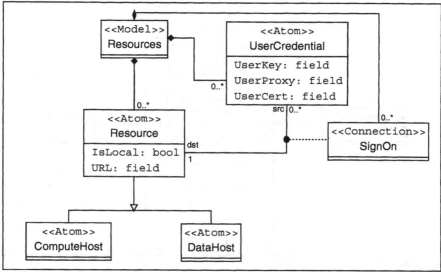

a) Resources

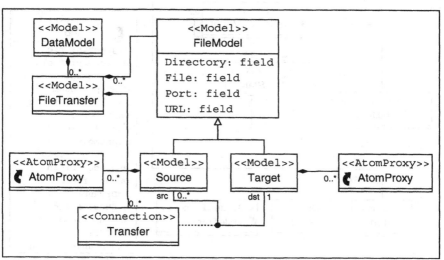

b) File Transfers

Figure 2.   Meta-model definition. The specification of the meta-model consists of four aspects: resources, file transfers, jobs, and workflows. (a) presents the definition of resources and (b) presents the definition of file transfers.

Using the meta-model, a user can define application workflows by interacting with the graphical environment provided by GME. The following example,

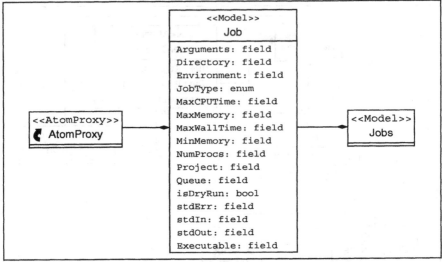

a) Jobs

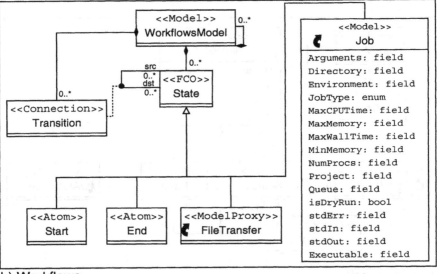

b) Workflows

Figure 3. Meta-model definition. The specification of the meta-model consists of four aspects: resources, file transfers, jobs, and workflows. (a) presents the definition of jobs and (b) presents the definition of the workflow part of the meta-model.

presents a simple application using Hidden Markov Models to illustrate how this interaction is performed for a typical Grid application. A Hidden Markov

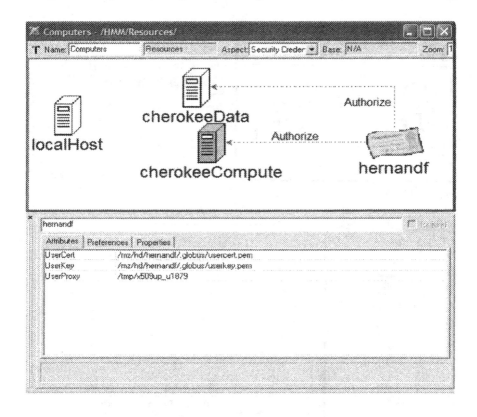

Figure 4. Model interaction. Once the meta-model is constructed, a user graphically defines the basic elements of the workflow. This figure shows how resources are defined and how a user credential authenticates the remote resources.

Model (HMM) was constructed to compare the differences between English and Spanish language patterns [11]. The input to the HMM is an intermingled file (parts in English and parts in Spanish) that only indicates if a letter is a vowel or a consonant (1 or 0). The output file consists of the language prediction. The subdivision of this application into different components and its integration in a Grid environment is presented in the rest of this section.

The application can be subdivided into pre-processing, HMM, and post-processing tasks. The input file is copied from the local computer to a remote host, and after the execution of the application, the output file is copied back to the local computer. The first step in defining the application involves the definition of resources. Light machines are used for data storage, while dark machines are used for computation purposes. Each remote resource needs to be authenticated with the user's credential (Figure 4). The next step is to define

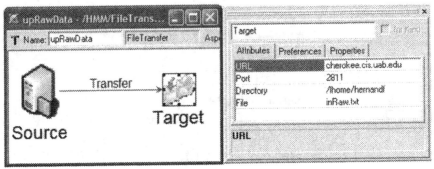

a) File Transfers

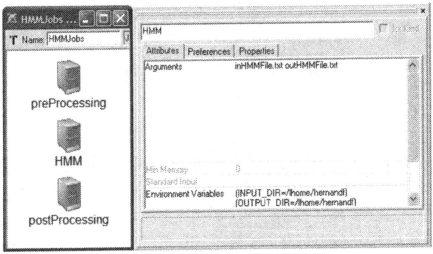

b) Jobs

Figure 5.   Model interaction. Once the meta-model is constructed, a user graphically defines the basic elements of the workflow: (a) shows that defining file transfers consist of specifying the location of the files on the endpoints; (b) shows how the HMM job is defined by specifying its RSL attributes.

the file transfers between the local computer and the remote host. In the case of uploading a file to the remote host, the URL and port of the host must be specified (Figure 5.a). Finally, the definition of jobs consists of the specification of the Resource Specification Language (RSL) attributes required to run the job (Figure 5.b).

After all of the tasks are defined, the application can be constructed by specifying the required sequence of tasks (Figure 6). File images indicate file transfers, and computer images indicate jobs to execute. The star in the far left

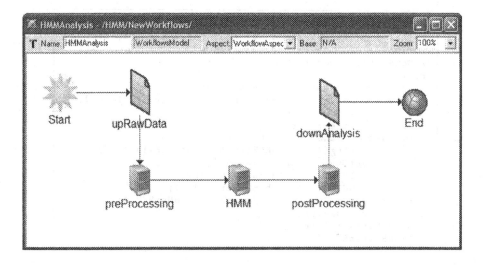

Figure 6.   Definition of the workflow in the model. This figure presents the workflow for the example. The input file is copied to the remote host (upRawData). A pre-processing job is executed on that file and its output is analyzed by the HMM job. The output of the HMM job is then modified in the post-processing step. Finally the output of the post-processing job is downloaded to the local computer (downAnalysis).

indicates the start of the workflow, and the sphere on the far right indicates the end of the workflow.

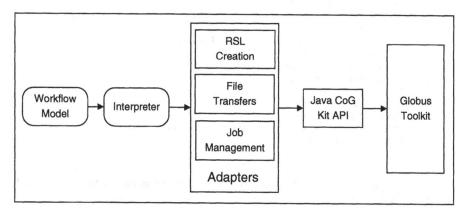

Figure 7.   Structure of the code generation. The interpreter traverses the workflow model and generates a Java program that interacts with a set of adapters. Creation of RSL strings, specification of file transfers, and remote execution of jobs are the facilities provided by the atapters. The adapters use the Java CoG Kit API to communicate with the Globus Toolkit.

```
1    //------------------------------------------------------------
2    // This code is generated by the interpreter when it
3    // finds a Job model in the workflow
4    // Generating Code for Job: HMM
5    try {
6        // load the user proxy certificate
7        byte[] hmmProxy = getByteArray("/tmp/x509up_u1879");
8
9        // create the rsl string
10       GlobusRSL hmmRSL = new GlobusRSL();
11       hmmRSL.setArg("HMM inHMMFile.txt outHMMFile.txt");
12       hmmRSL.setEnvironmentVariables(
13               "(INPUT_DIR=/lhome/hernandf)
14               (OUTPUT_DIR=/lhome/hernandf");
15       hmmRSL.setStdOut("/lhome/hernandf/stdOutHMM.txt");
16       hmmRSL.setNumProc(2);
17       hmmRSL.setDir("/usr/bin");
18       hmmRSL.setExec("java");
19
20       try {
21           GRAMJob hmmGRAM = new GRAMJob();
22           // submit the job
23           String hmmID =
24               hmmGRAM.submitJob("cherokee.cis.uab.edu",
25                                 hmmProxy, hmmRSL.toRSL());
26           // wait for its completion
27           String hmmCond =
28               hmmGRAM.checkStatusOfJob(hmmID, hmmProxy);
29           while (!hmmCond.equals("Job DONE."))
30               {
31                   System.out.println(hmmCond);
32                   Thread.sleep(1000);
33                   hmmCond =
34                       hmmGRAM.checkStatusOfJob(hmmID, hmmProxy);
35               }
36           System.out.println("hmmGRAM Done.");
37       } catch (Exception e) {
38           System.err.println(e.getMessage());
39           System.exit(-1);
40       }
41   } catch (Exception e) {
42       System.out.println
43       ("There was an error loading the user proxy.");
44       System.exit(-1);
45   }
```

Figure 8. Code generation. This code is generated by the interpreter for the HMM job submission task.

3.2   Interpreter

After the workflow is specified, a model interpreter traverses the internal representation of the model and generates the control code that manages the application execution. The interpreter first gathers all the information from resources, jobs, and file transfers. This information, along with the specification of the application workflow, constitutes the interpreter's input. The interpreter then executes the semantic actions associated with each workflow task. The output of this step is a Java program that manages the application execution. The Java program uses a set of supporting classes, or adapters [16], implemented to standardize the interaction with the Java CoG Kit API. The Java CoG Kit API is used as a bridge to communicate with the enabled back-end resources managed by the Globus Toolkit (Figure 7).

Figure 8 presents the code that is generated for the HMM job submission task. The process presented here is repeated for each job submission and similar code is generated. The references to the GlobusRSL on line 10 and the GRAMJob on line 21 are the adapters that communicate with the Java CoG API. Line 7 reads the user's proxy into a byte array. This is done to authenticate the user on the specific resource. Lines 10 through 18 create the corresponding RSL string. Finally, lines 21 through 36 submit the job to the specified resource and waits for its completion. As can be seen in this code, if some problem occurs then exceptions are caught and a corresponding message is displayed. File transfers operate in a similar way by using a different adapter.

4.   Related Work

The idea of composing applications from reusable components is not new. For example, WebFlow [1] introduces a platform-independent system that dynamically composes new applications from reusable components by clicking and dragging icons. The Job model of UNICORE [10] uses a set of directed acyclic graphs, and also permits the use of conditional and iterative execution of job groups or tasks. DAGMan [15] also maps a direct acyclic graph specification onto a physical environment. The Symphony framework [27] uses a graphical user interface for rapid collaborative development of grid applications following a data flow paradigm. Triana [30] also offers a visual programming model for the dynamic composition of predefined software components. Other works propose languages to specify Grid workflows. For example, Grid Workflow [6] focuses on proposing a standard for the sequence of complex high-performance computational tasks within a Grid. GridAnt [31] uses an XML-based language to specify client-side workflows. GridAnt is also able to submit the executions of tasks or file transfers by using a workflow engine based on the Apache ANT tool [5]. The construction of the workflow base aspect of the environment has been influenced by these projects.

Hategan et al., [4] proposes a technology and architecture-independent abstraction layer to provide interoperability across multiple Grid implementations, resulting in the Open Grid Computing Environment (OGCE). The main function of OGCE is to serve as a technology-independent, open, and extensible framework for client-side Grid development. This concept is similar to the idea presented in this chapter of using meta-models to abstract the underline Grid technologies. Because the models and the interpreter that translates those models are two different components of the modeling environment, with a change in the interpreter, the same models can be reused for different Grid architectures. This is an attempt to abstract the Grid environment into a high-level layer such that the essence of the workflow is not bound to a specific Grid environment. Furthermore, the abstractions provided by OGCE are comparable to those introduced in this chapter. For example, the task concepts presented in [4] contain concepts similar to those involved in the job specification (Figure 2). However, the main difference between the studies is in the level of abstraction. In this chapter the abstraction layer is realized at a domain-model level, but in [4] the abstraction layer is at a programming language level (Java).

5.   Future Directions

Work on the modeling environment is in its initial phase. The current implementation of the environment can handle only a limited number of sequential tasks in the workflow. At present, the generated applications communicate directly with the Java CoG Kit (as seen in Figure 8). This causes scalability problems due to the generation of specific code for each workflow task. A solution to this problem is currently under investigation and consists of developing a reusable workflow engine and generating appropriate configurations from the graphical models. In addition to improving the scalability of the generated applications, current efforts are aimed at three different areas: (1) allow parallelism of tasks in a workflow, (2) allow third party transfers, and (3) allow the definition of hierarchical workflows. A modified meta-model that considers these capabilities has already been implemented, but the corresponding interpreter is still in progress. Nevertheless, even without these capabilities, the initial experience with the environment is promising. In addition to this work in progress, future directions that will be considered involve four aspects:

1 In order to further simplify the use of the environment, integration with GIS [7] is the logical next step. This integration will provide feedback to users so they can decide which resources are more appropriate for their applications.

2 The current trend in Grid computing is moving towards a service architecture. To make the environment capable of moving in that direction, future work will be focused in two aspects: (1) the utilization of web services as workflow tasks, and (2) the capability of generating web services from workflows. The latter will allow non-web service applications to run and cooperate in a web service environment.

3 GridAnt [31] is another tool that allows a user to specify workflows for the Globus Toolkit. The difficulty posed by that tool is in the use of an XML file as an input. One of the major advantages of using the approach presented in this chapter is that more than one interpreter can be implemented for a particular meta-model. Because of this, the environment can serve as a front-end and by changing the interpreter, the required GridAnt's input file can be generated. The ability to generate multiple artifacts from the same model is a key benefit of model-driven techniques [26].

4 Currently there are different component frameworks that provide facilities to create Grid enabled applications [17, 2, 3]. Future work that is being considered involves the construction of model interpreters for these frameworks. Having these model interpreters will allow the composition of Grid applications incorporating components from different frameworks.

6.   Conclusion

The goal of the research described in this chapter is to improve the development of applications within the Globus Toolkit by creating graphical workflows of applications using domain-specific modeling techniques and the Java CoG Kit API. The benefits of using domain-specific modeling techniques which motivated this study were:

1 Domain modeling removes the accidental complexities of creating workflows in a Grid by focusing on higher levels of abstraction at the problem space *rather than solution space*, such as specific Grid libraries and their usage.

2 When exploring various workflows scenarios, modeling tools and their interpreters facilitate the more rapid ability to change the workflow details. That is, it is easier to manipulate and change domain models rather than the associated code.

3 Model-driven techniques possess the ability to generate multiple artifacts from the same model. Thus, with the same domain knowledge different output representations can be generated.

4 Domain models abstract a component model in an independent way, such that the specification in the models can be used with multiple component frameworks. The only requirement imposed by this methodology is the creation of a code generator for each particular component framework to be used. By creating different code generators (i.e., model interpreters),

components from different frameworks can be mixed and code for the corresponding framework can be generated.

Using these modeling techniques, a meta-model for the Globus Toolkit was created, as well as an interpreter that automatically generates Java code from the workflow models. With this approach, a programmer manipulates graphical models that represent the different components provided by the Globus Toolkit. From these models, the programmer is able to generate the corresponding Java programs that manage the execution of the application.

The potential impact of this study is the reduction of the development time involved in generating applications for the Globus Toolkit. Furthermore, users are not required to learn how to use the Java CoG Kit nor the Globus Toolkit to develop Grid-enabled applications. Rather, they construct graphical models that are at a more appropriate level of abstraction for describing the essence of the problem for a specific domain.

References

[1] E. Akarsu, F. Fox, W. Furmanski, and T. Haupt. WebFlow -High-Level Programming Environment and Visual Authoring Toolkit for High Performance Distributed Computing. In *Proceedings of the 1998 ACM/IEEE Conference on Supercomputing*, pages 1-7, San Jose, CA, 1998.

[2] M. Aldinucci, S. Campa, M. Coppola, M. Danelutto, D. Laforenza, D. Puppin, L. Scarponi, M. Vanneschi, and C. Zoccolo. Components for High Performance Grid Programming in Grid.it. In: *Component Models and Systems for Grid Applications*, pp. 19–38, Springer, 2004.

[3] M. Alt, J. Dünnweber, J. Müller, and S. Gorlatch. HOCs: Higher-Order Components for Grids. In: *Component Models and Systems for Grid Applications*, pp. 157–166, Springer, 2004.

[4] K. Amin, M. Hategan, G. von Laszewski, and N. Zulezec. Abstracting the Grid. In *Proceedings of the 12th Euromicro Conference on Parallel, Distributed and Network-Based Processing* (PDP 2004), pages 250-257, La Coruña, Spain, 2004.

[5] *The Apache ANT Project*. http://ant.apache.org/

[6] H.P. Bivens. *Grid Workflow*. GGF Grid Computing Environments Working Group Document, 2001. http://www.ggf.org

[7] K. Czajkowski, S. Fitzgerald, I. Foster, and C.Kesselman. Grid Information Services for Distributed Resource Sharing. In *Proceedings of the Tenth IEEE International Symposium on High-Performance Distributed Computing (HPDC-10)*, pages 181-184, IEEE Press, August 2001.

[8] K. Czajkowski, I. Foster, C. Karonis, C. Kesselman, S. Martin, W. Smith, and S. Tuecke. A Resource Management Architecture for Metacomputing Systems. In *Proceedings IPP-S/SPDP '98 Workshop on Job Scheduling Strategies for Parallel Processing*, pages 62-82, Springer-Verlag, 1998.

[9] E. Deelman, J. Blythe, Y. Gil, C. Kesselman, G. Mehta, K. Vahi, K. Blackburn, A. Lazzarini, A. Arbree, R. Cavanaugh, and S. Koranda. Mapping Abstract Complex Workflows onto Grid Environments. *Journal of Grid Computing*, 1(1):25-39, 2003.

[10] D.W. Erwin. UNICORE - a Grid Computing Environment. In *Concurrency and Computation: Practice and Experience*, 14(13-15):1395-1410, 2002.

[11] J. Fisher, F. Hernández, and A. Sprague. Language Patterns: Comparison and Prediction using Hidden Markov Models. In *Proceedings of the 41st Annual ACM Southeast Conference*, pages 246-250, ACM Press, Savannah, GA, 2003.

[12] *Flow Editor*. http://www-unix.globus.org/cog/projects/floweditor/

[13] I. Foster and C. Kesselman. Globus: A Metacomputing Infrastructure Toolkit. *The International Journal of Supercomputer Applications and High Performance Computing*, 11(2):115-128, 1997. http://www.globus.org

[14] G. Fox, D. Gannon, and M. Thomas. Overview of Grid Computing Environments. In *Grid Computing: Making the Global Infrastructure a Reality* (F. Berman, G. Fox, and T. Hey eds.), pages 543-554, John Wiley and Sons Ltd, Chichester, 2003.

[15] J. Frey, T. Tannenbaum, I. Foster, M. Livny, and S. Tuecke. Condor-G: A Computation Management Agent for Multi-Institutional Grids. *Cluster Computing*, 5(3):237-246, July 2002.

[16] E. Gamma, R. Helm, R. Johnson, and J. Vlissides. *Design Patterns: Elements of Reusable Object-Oriented Software*. Addison-Wesley, Reading, Massachusetts, 1995.

[17] D. Gannon, S. Krishnan, A. Slominski, G. Kandaswamy, and L. Fang. Building Applications from a Web Service based Component Architecture. In: *Component Models and Systems for Grid Applications*, pp 3–17, Springer, 2004.

[18] *The Globus Resource Specification Language RSL v1.0*. http://www.globus.org/gram/rsl_spec1.html

[19] J. Gosling, B. Joy, and G. Steele. *The Java language specification*. Addison-Wesley, 1996.

[20] J. Gray, T. Bapty, S. Neema, and J. Tuck. Handling Crosscutting Constraints in Domain-Specific Modeling. In *Communications of the ACM*, 44(10):87-93, October 2001.

[21] J. Gray, M. Rossi, J.P. Tolvanen. Preface: Special Issue on Domain-Specific Modeling. In *Journal of Visual Languages and Computing*, 15(3-4):207-209, June/August 2004.

[22] F. Hernandez. *Domain-Specific Models and the Globus Toolkit*. Tech. Rep. UABCIS-TR-2004-0504-1,Department of Computer and Information Sciences, University of Alabama at Birmingham, 2004. http://www.cis.uab.edu/info/grads/hernandf/papers/UABCIS-TR-2004-0504-1.pdf

[23] *JLex: A Lexical Analyzer Generator for Java(TM)*. http://www.cs.princeton.edu/~appel/modern/java/JLex/

[24] G. Karsai, M. Maroti, A. Lédeczi, J. Gray, and J. Sztipanovits. Composition and Cloning in Modeling and Meta-Modeling. In *IEEE Transactions on Control System Technology (special issue on Computer Automated Multi-Paradigm Modeling)*, 12(2):263-278, March 2004.

[25] B. Kernighan and D. Ritchie. *The C programming language*. Prentice Hall, 1988.

[26] A. Lédeczi, A. Bakay, M. Maroti, P. Volgyesi, G. Nordstrom, J. Sprinkle, G. and Karsai. Composing Domain-Specific Design Environments. *IEEE Computer*, 34(11):44-51, 2001.

[27] M. Lorch, and D. Kafura. Symphony - A Java-based Composition and Manipulation Framework for Computational Grids. In *Proceedings of the 2nd IEEE/ACM International Symposium on Cluster Computing and the Grid (CCGrid2002)*, pages 136-143, Berlin, Germany, 2002.

[28] J. Novotny. The Grid Portal Development Kit. *Concurrency and Computation: Practice and Experience*, 14(13-15):1129-1144, 2002.

[29] J. Sztipanovits and G. Karsai. Model-Integrated Computing. *IEEE Computer*, 30(4):110-112,1997.

[30] I. Taylor, M. Shields, I. Wang and O. Rana. Triana Applications within Grid Computing and Peer to Peer Environments, *Journal of Grid Computing*, 1(2):199–217, January 2003.

[31] G. von Laszewski, K. Amin, M. Hategan, N. Zaluzec, S. Hampton, and A. Rossi. GridAnt: A Client-Controllable Grid Workflow System. In *Proceedings of the 37th Hawaii International Conference on System Science*, pages 210-219, Island of Hawaii, Big Island, Jan 5-8, 2004.

[32] G. von Laszeweski, I. Foster, J. Gawor, and P. Lane. A Java Commodity Grid Toolkit. *Concurrency and Computation: Practice and Experience*, 13(8-9):643-662, 2001, http://www.cogkits.org/

[33] *Workflow in Grid Systems Workshop.* http://www.extreme.indiana.edu/groc/Worflow-call.html

[34] B. Yeo and B. Khoo. An Agent-based Grid Flow Management Framework for the Problem Solving Environment (PSE). In *Scientific Workflow Management Mini-Symposium*, GlobusWorld, 2004. http://www.globusworld.org/program/slides/1a_4.pdf

ON HIERARCHICAL, PARALLEL, AND DISTRIBUTED COMPONENTS FOR GRID PROGRAMMING

Françoise Baude, Denis Caromel, and Matthieu Morel
INRIA Sophia Antipolis, CNRS - I3S - University of Nice Sophia-Antipolis
Sophia-Antipolis – France
Francoise.Baude@sophia.inria.fr
Denis.Caromel@sophia.inria.fr
Matthieu.Morel@sophia.inria.fr

Abstract   We propose a parallel and distributed component model for building applications adapted to the hierarchical, highly distributed, highly heterogeneous nature of Grids. Instead of featuring a flat assembly model as for instance in the CCM and CCA models, we claim that a hierachical assembly model should ease the building and dynamic reconfiguration of component oriented Grid applications. The proposed model and associated framework is based on ProActive, a middleware (programming model and environment) for object oriented parallel, mobile, and distributed computing. We have extended ProActive by implementing a hierarchical and dynamic component model, named Fractal, so as to master the complexity and scalability of composition and deployment. This defines a concept of components for the Grid: primitive or composite, made of several activities, parallel and distributed. Components communicate using typed one-to-one or collective invocations on interfaces. Composition of interfaces and of other properties such as the one pertaining to the deployment of components are specifically addressed.

Keywords: active objects, hierarchical components, deployment, dynamic configuration, group communications

1.   Introduction

In order to motivate our research, we begin by recall what are the requirements of Grid programming models, and then, what are the specificities of the other research on component oriented models for Grid programming with which our work can be compared.

1.1   Context and Related Work

Component programming for Grid and peer-to-peer computing is gaining growing interest, as it is considered of being capable to tackle the complexity, dynamicity and heterogeneity of target applications and their support maybe more easily than other approaches. Examples of alternative programming models currently in use include MPI for message passing, and GridRPC [22] for remote procedure calls. Indeed, one can consider Grid programming as requiring a two-level programming approach [12]: nuggets or code modules are generated by conventional programming, that must be augmented for the Grid by the integration of the distributed nuggets together into a complete executable. Of course, each nugget may be something as complex as a parallel and distributed application in itself. The user can be offered many different paradigms for expressing this integration. One common model is a graphical interface where nuggets are chosen from a palette and linked via their ports or channels. A perhaps more powerful way is to program the linkage, via scripting or compiled programming languages. Whatever the paradigm, the idea is first to wrap services (applicative or even Grid services) as components and second to rely on a framework so as to instantiate those components and to allow them to be composed into applications. As the time a component instance's *server* interface (server interfaces provide services) is invoked, the appropriate actions are taken by the component implementation. Those actions may involve `client` interfaces (client interfaces require services) of linked component instances.

Examples of this include the CCA model [15], which defines components, and where instantiation and composition are implemented within a framework (for instance, XCAT [5, 18]); the ICENI project [14]; the GridCCM project [11], which relies on the Corba Component Model for the component definition and whose specificity is to efficiently embed parallel MPI codes. One important remark is that all of the known component models for Grid programming enable an assembly of components which is only *flat*. This chapter proposes a novel approach through which Grid applications will be build by assembling components in a *hierarchical* way instead.

1.2   Contribution

Our claim regarding component-oriented Grid programming is as follows:

1 a set of components, assembled or not, may usefully yield to a new *composite* component that can recursively be composed with other components. This is a way to enforce code reuse and scalability of the composition task, because it structures this task. More precisely, it enables the user which builds an application by composition, to naturally proceeds in a hierarchical and structured manner. With such an approach, we aim at easing the programming, the deployment, and eventually the monitoring of complex Grid applications;

2 inclusion, bindings, and also location of components, must be *reconfigurable*. This is a way to adapt to the dynamicity of Grid runtime environments. For instance, if a component fails, it is possible to replace it by a new instance, then to rebind its enclosing component so as it references it, without any other consequence for the rest of the application.

As a concrete illustration of those ideas, we describe the first version of a component model for Grid computing we have defined and implemented within the ProActive middleware [3]. The specification of the components is conformable to the Fractal component model [6, 13], a generic and extensible software composition framework. We provide an implementation of the Fractal specification API within the ProActive library [21] for parallel, distributed computing [2, 8]. The library is based on an active object pattern that is a uniform way to encapsulate: a remotely accessible object, a thread as an asynchronous activity, an actor with its own script, a server of incoming requests, a mobile and potentially secure agent.

In Section 2, we give a brief but complete overview of the proposed component model; in Section 3, we detail two specific points that must be tackled with when targeting Grid computing: distributed mapping and parallelism.

2. Overview of ProActive Components

Primitive components. By implementing the Fractal component model within the ProActive library, components at the *primitive* level, are themselves formed of one or several active objects (i.e. a primitive component is a nugget that may be parallel and distributed). Standard meta-information (e.g. XML) technique has to be used for identifying provided and used ports or interfaces. Those ports are typed, and no IDL is required as the components are all defined using the Java language. Currently, we are working on the design of a generic wrapper implemented as active objects, whose aim is to encapsulate legacy parallel codes (i.e. Fortran-MPI, C-MPI codes). But the Grid specificity calls for specific information related to the parallel and distributed nature of codes. More importantly, an abstraction for the mapping of such codes must be included in the component meta-information for the deployment. Such an abstraction is named a *virtual node* [2]. Secondly, the client and server interface specifications must authorize collective behavior. Collective communications to implement parallel propagation of service invocations are essential for efficiency purposes. Such a collective method invocation feature between active objects is available in the ProActive library [1]; this feature is used for implementing *collective ports* or interfaces. Other properties such as quality of service requirements for instance may be identified in the future.

Composite components.   Primitive components being defined, the next step to master complexity and scale of Grid applications is to be able to compose those building blocks into new components called *composite*. The resulting encapsulated composite component can be seen as a functional and autonomous subsystem. Recursively, a composite component can be defined as the composition of primitive or composite components (see Figure 1). Components systems can be initially described in a declarative manner using an architecture description language (ADL) [9], where are specified, in a standardized XML format, components definitions, assemblies and bindings. An important aspect concerns the recursive composition of specific attributes with respect to Grid defined at the level of inner components. Composing virtual nodes is the opportunity to decide colocation of components or on the contrary to decide remoteness. Composing two ports, when at least one of them is a collective one is the mean to couple parallel codes via synchronization and data redistribution: e.g. N-to-1, 1-to-N or N-to-M communication patterns.

Parallel composite components.   A specialization of composite components is *parallel* components: all the inner components of the composite component are of the same component type; invoking a provided interface on the parallel component triggers its parallel propagation to all the inner components.

A component is deployed and run through the container that ProActive transparently offers via a set of meta-level objects. In particular, meta-level objects implement all the component controllers specified by the Fractal model (more importantly, the `binding`, `content` and `life-cycle` controllers). This implies that any component whatever its nature, is associated to – at least – one active object, the one which references meta-level objects.

As ProActive implements the Fractal specification (except component templates and sharing), it is possible to dynamically start/stop component's life-cycle, reconfigure bindings and inclusions (by invocation of the Fracal API methods, such as `unbindFC`, `bindFC`). This is thus the way to modify the initial description of a component system, incorporate newly created or discovered components, reconfigure assemblies, etc.

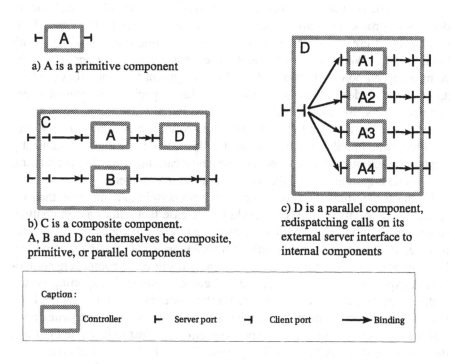

a) A is a primitive component

b) C is a composite component.
A, B and D can themselves be composite,
primitive, or parallel components

c) D is a parallel component,
redispatching calls on its
external server interface to
internal components

Figure 1.   The three types of components

3.   Specific Properties of Grid Components

We will now first detail how the deployment of distributed and hierachical components on Grid infrastructures can be managed, and second, how to compose collective interfaces of components.

3.1   Distributed Deployment

We add an important property to the definition of a component for the Grid: the information on its distribution, not in the form of concrete location, but instead in the form of one or several *virtual nodes*. This way, only virtual node names – not statically defined physical locations – appear in the component definition. It is the component framework executing the deployment of the application on the target architecture that maps virtual node names (i.e. those symbolic names) to concrete physical locations. In the ProActive model, physical locations are represented in the form of handles/entry points to corresponding created or acquired remote JVMs.

Mapping of a virtual node to a concrete JVM is described within a deployment descriptor: protocol to start a process on a remote host (ssh, globus Gram, LSF, etc), so as to create a JVM with a ProActive runtime (need path, classpath for instance). We assume that it is the user that builds up an application by composing components that provide such a deployment descriptor for the whole application. Indeed, each application may have specific deployment needs and parameters, so its complete deployment is initially defined in a global and coherent way in such a deployment descriptor. As we run the components within the ProActive middleware, we automatically provide a way to dynamically migrate the activities of the components on new locations, while they are running (this is one important part of the reconfiguration requirement).

Composing virtual nodes of two components pertain to composing their respective sets of virtual nodes. It might be the case that names are in conflict, so the composition may superimpose renaming. The superposition must be remembered in the composite definition for further usage at the deployment time. For instance, assume a hierarchical component resulting of composing component A exporting virtual node named *vn1* and component B exporting virtual node also named *vn1*. A design choice for the hierarchical component might be to co-allocate both components on the same JVM. In this case, the virtual node that the hierarchical component exports/specifies has just to be named *vn1* and the deployment proceeds normally by instantiating the hierarchical component and both inner components on the JVM associated to *vn1*. On the contrary, the user may want to be sure that the two inner components run on different JVMs. In this case, during the composition, he renames both virtual names as, for instance, *vn1-A* and *vn1-B* and specifies the set {*vn1-A, vn1-B*} as the virtual node property value for the hierarchical component. Further, he defines the deployment of *vn1-A* and *vn1-B* onto two distinct JVMs in the deployment descriptor. The hierarchical component's active object is indifferently created on *vn1-A* or *vn1-B* without a specific mention of the user.

3.2   Collective Communication

As the proposed model targets high-performance Grid computing, there is an additional need compared to the original Fractal component model: parallelism. This means first, expressing parallelism at the component level and secondly, be able to implement it efficiently.

Requirements.   The aim is to provide the user which builds up an application, a very simple way to identify and structure components that should run in parallel and as such, be mapped on distinct computing resources. This also requires that the forwarding of a service call on a group of components be itself as parallel as possible, so that the services really have a chance to run in parallel.

Solution. We provide a way to group services into groups of services of the same type. We use the notion of *collective port*. This implies the following: when calling a service on this port, the service is propagated to the group of components it is bound to. As a programming-level constraint, only a client port defined as collective can be safely bound to a group of components. Also, the component model has to provide a way to get the current number of instances in the collective port, initially and during execution, as this number of instances may dynamically change (this is one important requirement for the reconfiguration aspect).

Implementation. We rely on ProActive's group communication mechanism which achieves asynchronous remote method invocation for a group of remote active objects of the same type, with automatic gathering of replies [1] (other but similar mechanisms for collective remote method invocations towards distributed objects exist, e.g. [10, 19]). Recall that each component is implemented by one active object (even composite components). A group of services, that can also be seen as a `java.util.Collection`, is simply represented as a group of the associated references to active objects, and a collective port is indeed a proxy-stub to the group. Moreover, a method call to the group is optimized compared to the sequential achievement of individual calls (i.e. multi-threaded sending of all calls, only one serialization phase for the parameters; note that by default, each parameter is *broadcast*ed to each individual call). When invoking a service on a collective port, an important specificity (resulting from the Proactive group communication mechanism) is that the result group is also a group, transparently built at invocation time, with a future for each elementary reply. The result group is immediately returned to the caller. It will be dynamically updated with the incoming results, thus *gather*ing results. It is possible to slightly modify the semantics of a service invocation through a collective port : in case a parameter is a group of values, each value can be asked to be *scatter*ed to each individual call in a round-robin fashion.

Usage of a collective port is thus the way for a component to be bound to a group of components. There are in fact at least two situations where the usage of a collective port arises: as a client port, and as a server internal port of a parallel component. These two cases are explained in more details below:

Collective client port. A client port can be specified as being a collective one (see Figure 2b). In this case, the component implementation (on the figure, the implementation of the component on the left) is prepared so as to take advantage of a collective call on such a client port. Among others, it might have programmed how to synchronize – and further *combine/reduce* – on the results that will come back from each individual call (waitAll, waitAny, wait(result number i), etc). It is of course completely dependent upon the component implementation. In case the implementation is not prepared to be returned a group of result instead of a single result, there might yield to a programming error. For this reason, binding a group of components to a client port requires that this port be defined as being collective. What is the benefit of exposing this sort of client port at the component level? The user which builds up or use such an application is aware of the fact that he can statically and dynamically adapt the number of instances bound to this collective port to his needs (for ease of structure, performance or fault-tolerance reasons).

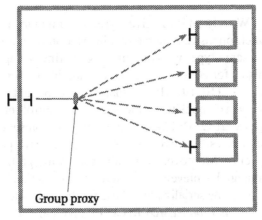

Group proxy

a) group communication inside a parallel component : incoming calls on the server interface are dispatched to the inner components.

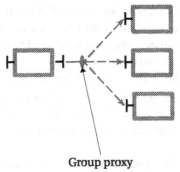

Group proxy

b) group communication as the implementation of a collective client port

Figure 2.   Group communications allowing collective bindings and parallel components

A special case for a composite component: a parallel component. It is also a composite, but re-dispatching calls to its external server interfaces towards its inner components (see Figure 2a), with all parameters broadcasted (default behavior). The number of parallel inner components that must be instantiated at deployment time is defined externally in the deployment descriptor: by default, it is equal to the number of different JVMs that are created during the activation of the virtual node specified by the parallel component. Alternatively, the parallel component may have set a specific attribute which gives the required number of instances. The bindings between each external server port of the parallel component to the corresponding server ports of each instance are performed transparently and automatically inside the parallel component (the instances are created as a ProActive group): for this reason we can say that an internal server port of a parallel component is a *collective server port*.

Composing collective ports. It is what we plan to explore in order to couple parallel components. Every parallel component has an active object associated to it and so could serve as a sophisticated re-dispatcher: for the set formed from each client port of interest of each inner component of the parallel component (thus defining the notion of a collective client port of a parallel component), to the server collective port of interest it is bound to. The objective – but maybe not the solution – is similar to what is achieved by introducing collective communications as tees in the ICENI Grid oriented component model [20]: switch, combiner, splitter, gather, broadcast; the same regarding the collective port extending CCA ports, experimented in [17], which is in fact implemented as a combination of translation components (i.e. customizable components, efficiently called by the framework, to tackle translation/redistribution of data, collective invocation and returns, e.g. a MxN component). Our challenge is to provide a solution adapted to the component-oriented model we propose, that is, without the explicit introduction of additional components (either generic or programmer-modifiable), but only through the definition of Fractal ports and the usage of the ProActive group communication mechanism. In this sense, our design of collective ports composition should end up quite close to the one which is based on collective RMI calls extended with the usage of a MxN redistribution scheme introduced in recent CCA compliant implementations of parallel data redistribution [4, 10].

4. Conclusion

We have successfully defined and implemented a component framework for ProActive, by applying the Fractal component model, mainly taking advantage of its hierarchical approach to component programming. This defines a concept of what we have called *Grid components*. Grid components are formed of parallel and distributed active objects, features mobility, typed one-to-one or collective service invocations and a flexible deployment model. They also enables flexibility, dynamicity, structure and scalability at the composition task level. Not all component models are equivalent. The one we think is well adapted for complex softwares such as those for the Grids, is not based on a flat assembly model, but on a hierarchical one (even if this induce some overhead due to the presence of those 'additional' components which are the composite ones). Moreover, it features reconfiguration capabilities. This means dynamically be able to change bindings, inclusion, location. Reconfiguration is the next big research issue for which we are currently considering the following points:

- have functional method calls (either single or collective) bypass each inner composite component of a hierarchical component, so as to directly reach target primitive components; but at the same time, enable of coherent re-binding even in presence of direct functional communications

- for the sake of reliability and fault tolerance which is crucial when components are applied to Grid computing – or even worse in case of peer-to-peer global computing–, error and exception handling across components, checkpointing, volatility, and security must be considered.

We think that these features will be key-points for reaching adaptive Grid middleware, with dynamic strategies at various points (for communications, checkpointing, reconfiguration, . . .) in presence of various conditions (SAN, LAN, WAN, P2P, etc).

We are also working on GUI-based tools to help the end-user manipulate Grid component based applications. Those tools will extend the IC2D monitor [2] provided in the ProActive middleware, which already helps in dynamically changing the deployment defined by deployment descriptors (acquire new JVMs, drag-and-drop active objects on the Grid): we intend to provide *graphical interactive* dynamic manipulation and monitoring of the components (besides what can already be done by programming once the application has been deployed).

Such tools could be integrated to computing portals or with Grid infrastructure middleware for resource brokering (as done for instance in ICENI [14], XCAT [18], etc.), so as to build dedicated Problem Solving Environments (see SciRun for instance [16]). Compared to those previous frameworks which relies on a flat assembly model of components for describing applications, the proposed component model is a more general one as it allows structured/hierarchical composition. As such, the hierarchical and structured dynamic manipulation of component-based applications, through portals or any other means, becomes naturally possible.

Moreover, the underlying activities of ProActive components are based on an underlying computing model which follows a sound semantics [7]. So, there is opportunity to construct well-defined behavioural semantics for components and for their assemblies.

References

[1] L. Baduel, F. Baude, and D. Caromel. Efficient, Flexible, and Typed Group Communications in Java. In *Joint ACM Java Grande - ISCOPE 2002 Conference*, pages 28–36, ACM Press, Seattle, 2002.

[2] F. Baude, D. Caromel, F. Huet, L. Mestre, and J. Vayssière. Interactive and Descriptor-based Deployment of Object-Oriented Grid Applications. In *11th IEEE International Symposium on High Performance Distributed Computing*, pages 93–102, 2002.

[3] F. Baude, D. Caromel, and M. Morel. From Distributed Objects to Hierarchical Grid Components. In *CoopIS/DOA/ODBASE*, LNCS, pages 1226–1242, Springer, 2003.

[4] F. Bertrand and R. Bramley. DCA: A Distributed CCA framework based on MPI. In *9th International Workshop on High-Level Parallel Programming Models and Supportive Environments at IPDPS*, April 2004.

[5] R. Bramley, K. Chin, D. Gannon, M. Govindaraju, N. Mukhi, B. Temko, and M. Yochuri. A Component-Based Services Architecture for Building Distributed Applications. In *9th IEEE International Symposium on High Performance Distributed Computing Conference*, 2000.

[6] E. Bruneton, T. Coupaye, and J. Stefani. Recursive and Dynamic Software Composition with Sharing. *Proceedings of the 7th ECOOP International Workshop on Component-Oriented Programming (WCOP'02)*, June 2002.

[7] D. Caromel, L. Henrio, and B. Serpette. Asynchronous and Deterministic Objects. In *Proceedings of the 31st ACM Symposium on Principles of Programming Languages*, ACM Press, 2004.

[8] D. Caromel, W. Klauser, and J. Vayssiere. Towards Seamless Computing and Metacomputing in Java. *Concurrency Practice and Experience*, 10(11–13):1043–1061, November 1998.

[9] P.C. Clements. A Survey of Architecture Description Languages. In *International Workshop on Software Specification and Design (IWSSD'96)*, pages 16–25, 1996.

[10] K. Damevski and S.G. Parker. Parallel Remote Method Invocation and M-by-N Data Redistribution. In *4th Los Alamos Computer Science Institute Symposium*, 2003.

[11] A. Denis, C. Pérez, T. Priol, and A. Ribes. Bringing High Performance to the CORBA Component Model. In *SIAM Conference on Parallel Processing for Scientific Computing*, February 2004.

[12] G. Fox, M. Pierce, D. Gannon, and M. Thomas. Overview of Grid Computing Environments. Technical report, Global Grid Forum document, 2003. http://www.ggf.org/documents/final.htm.

[13] Fractal. http://fractal.objectweb.org.

[14] N. Furmento, A. Mayer, S. McGough, S. Newhouse, T. Field, and J. Darlington. ICENI: Optimisation of Component Applications within a Grid Environment. *Parallel Computing*, 28(12), 2002.

[15] D. Gannon, R. Bramley, G. Fox, S. Smallen, A. Rossi, R. Ananthakrishnan, F. Bertrand, K. Chiu, M. Farrellee, M. Govindaraju, S. Krishnan, L. Ramakrishnan, Y. Simmhan, A. Slominski, Y. Ma, C. Olariu, and N. Rey-Cenvaz. Programming the Grid: Distributed

Software Components, P2P and Grid Web Services for Scientific Applications. *Cluster Computing*, 5(3), 2002.

[16] C.R. Johnson, S. Parker, D. Weinstein, and S. Heffernan. Component-based Problem Solving Environments for Large-scale Scientific Computing. *Journal on Concurrency and Computation: Practice and Experience*, (14):1337–1349, 2002.

[17] K. Keahey, P. Fasel, and S. Mniszewski. PAWS: Collective Interactions and Data Transfers, In *Proceedings of the IEEE International Symposium on High Performance Distributed Computing (HPDC'10)*, 2001.

[18] S. Krishnan and D. Gannon. XCAT3: A Framework for CCA Components as OGSA Services. In *9th International Workshop on High-Level Parallel Programming Models and Supportive Environments (HIPS)*, 2004.

[19] J. Maassen, T. Kielmann, and H.E. Bal. GMI: Flexible and Efficient Group Method Invocation for Parallel Programming. In *LCR'02: Sixth Workshop on Languages, Compilers and Run-time Systems for Scalable Computers*, 2002.

[20] A. Mayer, S. Mcough, M. Gulamali, L. Young, J. Stanton, S. Newhouse, and J. Darlington. Meaning and Behaviour in Grid Oriented Components. In *Third International Workshop on Grid Computing, GRID*, LNCS, 2536:100–111, 2002.

[21] *ProActive Web Site*. http://www.inria.fr/oasis/ProActive/.

[22] K. Seymour, H. Nakada, S. Matsuoka, J. Dongarra, C. Lee, and H. Casanova. GridRPC: A Remote Procedure Call API for Grid Computing. In *Third International Workshop on Grid Computing, GRID*, LNCS, 2536:274, 2002.

ICENI: AN INTEGRATED GRID MIDDLEWARE TO SUPPORT E-SCIENCE

Anthony Mayer, Steve McGough, Nathalie Furmento, Jeremy Cohen, Murtaza Gulamali, Laurie Young, Ali Afzal, Steven Newhouse, and John Darlington
London e-Science Centre
Imperial College London
South Kensington Campus
London, UK
lesc-staff@doc.ic.ac.uk

Abstract   Scientists now have a greater desire to undertake work within global collaborations. This increases their dependence on distributed computation, storage and data resources. For this new paradigm of e-research to be easily adopted by the applied science community, it needs to be enabled by a new software infrastructure – Grid middleware. In this chapter, we describe ICENI, an integrated Grid middleware that explores the services and meta-data necessary to support e-research within a variety of application domains. We focus on the services that we feel are necessary to facilitate Grid use, ranging from running a simple self contained application through to building a simulation from scientific software components distributed across a Grid, selecting the optimal combination of services to enact the simulation and paying for them on demand.

Keywords: Grid middleware, OGSA, component programming model, e-Science, scheduling, performance, advance reservations, meta data

1.   Introduction

ICENI (Imperial College e-Science Networked Infrastructure) has originated from the research activities of Professor John Darlington and colleagues in the 70s and early 80s in the development and exploitation of functional languages. The growth of applied parallel computing activities at Imperial College demonstrated a fundamental need for a software environment to enable the use of complex resources by the average scientist. This requirement became even more apparent with the growth and adoption of Grid computing within the UK (a significantly more complex environment than a single parallel machine) to enable computer based research – e-research. The enduring goal of ICENI is to increase the effectiveness and applicability of high-performance methods and infrastructure across a whole range of application areas in science, engineering, medicine, industry, commerce, and society.

Our focus within ICENI therefore has three major elements: prototyping the services and their interfaces necessary to build a service oriented Grid middleware, developing an augmented component programming model to support Grid applications, and to explore the meta-data needed to annotate the services and software to enable effective decision making about component placement within a Grid.

In this chapter, we present an overview of the ICENI Grid middleware describing many of the recent developments within this architecture. The aim is to show how drawing these new technologies together provides an improved user experience of the Grid. Details of each of these new technologies may be obtained from the papers referenced in the appropriate section. We exemplify the 'real world' use of ICENI through two of the external projects in which we are involved, we further speculate as to the 'added value' our new features will provide for users of these, and other, projects.

The rest of the chapter is organised as follows. Section 2 presents the overall ICENI architecture, more details on the architecture can be found in [7]. The component programming model is presented in Section 3, more details can be found in [16–17]. Sections 4 and 5 present some of the higher level services we are developing to support our current application communities. The scheduling services are presented with more details in [1, 6, 28] and the economic services in [5, 14]. The aim of these sections is also to show how these different aspects of the ICENI middleware are brought together to provide enhanced solutions to the user's requirement. This is demonstrated in Section 6 through two of the external projects we are involved in. Section 7 presents comparisons with similar activities before Section 8 concludes and outlines some elements of the future ICENI roadmap.

2. Architecture Overview

This section presents an overview of the service oriented architecture that is the basis of the ICENI middleware, as well as the mechanisms provided for the deployment of ICENI.

2.1 The Service Oriented Architecture

The ICENI architecture, which is developed in Java, is an example of a service oriented Grid middleware that is able to use different technologies (such as Jini [11], Web services [27] or peer-to-peer infrastructures such as JXTA [12]) to instantiate the service architecture [7]. ICENI provides implementation independent abstractions that allow a service provider to develop and advertise a service into the service registry where it may be discovered by other members of the community – clients searching for a particular service type. On finding an appropriate service, the client retrieves information from the registry that will describe how the service instance may be contacted and used.

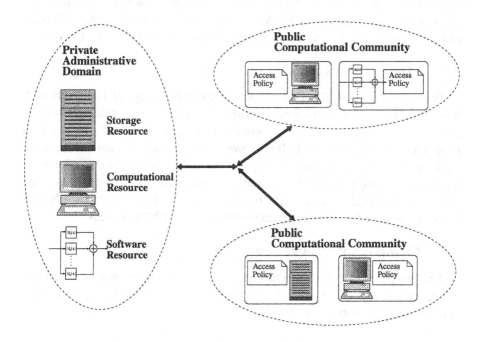

Figure 1. The ICENI service oriented architecture

Within ICENI, we use the Service Oriented Architecture (SOA) infrastructure to provide a mechanism for federating the resources managed by a real organisation into a pool of services that may be used by a virtual organisation or a computational community (see Figure 1). There are two key abstractions:

1. a resource in a private administrative domain represents a capability owned by a real organisation that may be exposed through ICENI for sharing with a virtual organisation;

2. a service in the public computational community is a resource that has been annotated with a policy that describes when and by whom the resource may be used.

The implementation neutral SOA interfaces within ICENI services may be implemented through Jini (currently), JXTA (under development), Web Services, or a text based registry such as LDAP [21]. This decoupling between the services and the implementation that hosts them allows us to focus our efforts on the higher level services needed to build an Open Grid Services Architecture [19] (OGSA) and the support needed from the infrastructure rather than the specific details of the infrastructure itself. An alternative approach is to expose these services to other infrastructures through 'gateways'. We have exposed the Jini services within ICENI through an OGSI [20] (Open Grid Service Infrastructure) gateway enabling interoperability between the two systems.

2.2   Deployable Grids

The deployment of ICENI is simplified through the use of Java Web Start based installation and configuration wizards. Java Web Start [25] allows a cross-platform wizard based installation process to be launched directly from a Web page. Once the necessary ICENI files have been downloaded and installed, the configuration wizard provides a simple point-and-click interface to take users through a step-by-step process to configure and launch ICENI services. The wizard offers a quick and easy way for new users to get started with ICENI and is also accessible for subsequent configuration of ICENI resources.

3.   Component Programming Model

The service oriented architecture and the component programming model are the two assets of the ICENI Grid middleware. This section presents the component programming model.

3.1   Layered Meta-Data to Enable Component Programming

Grid based programming requires a high-level programming model that performs in a resource / platform independent fashion. While some modern languages (e.g. Java) provide supposedly platform independent programming, the e-Science context requires the ability to deploy high performance codes that have been compiled specifically for certain hardware configurations. The solution to the tension between high performance and portability is to raise the level of abstraction. Where one possesses multiple implementations of a specified behavioural pattern, that pattern can be deployed on whatever hardware resources are available. For this reason, ICENI defines a component programming model with distinct layers of meta-data describing the component's meaning, behaviour and implementation [17].

An end-user who wishes to assemble an application uses tools to discover ICENI software resources which appear as services within the architecture. These resources provide the component meta-data for available components. The user then composes the application in terms of meta-data descriptions as the highest level of abstraction: the level of meaning. Once composed this application can be reused whenever any software resources exist that support the meanings expressed in the composition. Each meaning may have multiple behaviours, which express the internal workflow model of the component, and each behaviour may represent multiple implementations, where performance characteristics and hardware resource requirements are specified.

3.2   Spatial and Temporal Expressions

The design choices that motivate the separation of concerns between the meta-data layers are driven by the distinction between spatial and temporal expressions of composition. The application is composed using 'spatial' relationships, where there is no temporal ordering in the representation. As such, components do not represent dependent activities in a task graph, but rather concurrently existing actors communicating through channels. We adopt this model as it allows interactive modifications to composition at run-time: to give an example, consider an existing executing composition consisting of a simulation and a visualisation component. These can be discovered dynamically, and a steering component added to the composition without any redefinition of the existing relationship between the simulation and visualisation. Adding and removing components from the workflow is made substantially easier with a non-temporal representation, and is a key feature of the ICENI component programming model [16].

The disadvantage of using a spatial workflow model is that the performance modelling mechanism requires a task graph of the composed application's activities – this is where the behavioural layer is brought into play. The behavioural meta-data specifies the task graph of activities that occur upon the invocation of each port on a component. By taking this into account alongside the spatial composition of components, it is possible to assemble a composite task graph for the entire application. This graph is then used to create performance models for each possible set of implementations of the available components.

3.3   Implementation Selection & Scheduling

The temporal expression of workflow, the task graph, can be labelled with performance data for activities being executed on distinct hardware resources. From this graph, it's possible to build sets of composite performance models for each possible allocation of resource/implementation combination and compare them. This information can be utilised by the scheduler, which may not necessarily always select the fastest resources (if, for example, it is also accounting for their cost). We describe the exploitation of the performance information in the next section.

4.   Performance Driven Scheduling & Reservation

This section presents the scheduling mechanisms provided by ICENI. These mechanisms are built on top of the service oriented architecture and the component programming model presented in the two previous sections.

4.1   Overview

The meta-data provided by the component programming model is now used to drive an intrinsic trinity between the scheduling of a workflow, its behavioural and performance meta-data (available about the components within the workflow) and the ability to reserve resources [1, 6]. Without performance modelling one cannot extract a critical path, nor produce accurate reservations, and without reservations one cannot control the predictability of execution times. In this section, we outline the scheduling and performance architecture developed within ICENI to exploit this trinity.

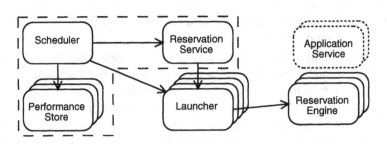

Figure 2.   The trinity architecture: scheduling, reservation and performance prediction

The basic trinity architecture is illustrated in Figure 2. A Scheduler may interact with multiple Launchers (Job Management Services) representing one or more resources. A Launcher may be associated with one Reservation Engine, if reservations are possible. The Scheduler interacts with the Reservation Service which in turn communicates with the Reservation Engines through the appropriate Launcher. There may be multiple Performance Stores each of which can be interrogated by the Scheduler. Once a workflow is instantiated it will have an Application Service which exists until the workflow terminates.

4.2 Scheduling

Components within a workflow are scheduled, through the ICENI scheduling system which determines the 'best' mapping of component implementations onto a subset of available resources through the goals of the various stakeholders i.e. users, resource providers and managers of the virtual organisation. Data from the Performance Repository, meta-data provided by the component implementor and meta-data from the resource owner can be used in determining the 'best' mapping. With this information, the scheduler can determine the critical path and produce (potentially more than one) concrete workflow using a subset of the resources available within the Grid. These concrete workflows can be passed to the Reservation Service to generate reservations. Once a single concrete workflow has been selected and resources reserved, the scheduler will instantiate the planned workflow and expose the running components through an Application Service.

A number of scheduling algorithms are available to perform this task, such as simulated annealing, complete information game theory, best of N random and exhaustive search [28]. These algorithms can be used within the ICENI environment using the pluggable scheduling framework. Each of these algorithms at some point has to evaluate and contrast potential schedules. When optimising for time (as in most cases), this is done by computing the expected execution time. By using the task dependency graph and the performance meta-data, the expected start and end times for each activity are evaluated and the scheduler can then select concrete workflows with the shortest overall duration. This possible execution plan is passed to the Reservation Engine to obtain dedicated reserved resources to support its execution.

4.3 Performance

The Performance Repository system developed within ICENI is capable of monitoring running applications to obtain performance data for the components within the application. This may range from basic start and stop times to more complex metrics such as operation count. This data is stored within the Performance Repository with meta-data about the resource that executed the component, the component implementation and the number of other components concurrently running on the same resource. There is also provision for the component implementation designer to define other meta-data that should be stored. This could include such things as the problem characteristics which will affect the execution time.

In future runs of the ICENI system, the performance data stored can be used by the scheduler to estimate the execution times for each component within the workflow and hence the overall execution time of the application. The rich meta-data stored in the Performance Repository can be used to tailor the returned data to more accurately reflect the current circumstances.

4.4   Reservations

The Reservation Service attempts to co-allocate all the resources required for the execution of the workflow. It does so by entering into negotiation with the appropriate resource manager, abstracted through the Reservation Engine, running on each resource in order to make advanced reservations for the time and duration specified in the concrete workflow. The negotiation is bound by the constraints defined by the earliest and latest start times for an activity as calculated by the scheduler. The negotiation protocol is based on WS-Agreement [9].

The Reservation Engine handles all incoming reservation requests. These requests are validated in terms of the user access permissions and the availability of the named resource. Then the Reservation Engine checks for conflicts with existing reservations. If a conflict is detected, i.e. the time frame for which a reservation is requested overlaps that of an existing reservation on the same resource, the reservation request is rejected and the client notified. If the reservation request is valid and there are no conflicts, the Reservation Engine creates a reservation on the resource. The Reservation Engine can also potentially return a list of alternative reservations: reservations on the same resource, for the same duration of time, but beginning at a different time. The client can then choose to create a reservation from the list of alternatives presented to it or abandon the negotiation process.

5.   Economic Services

With transparent and composable services being made available through the ICENI system, the possibility emerges of creating a true market in computational services. By adding the ability to account and charge for service usage, an open market is envisaged, in which resources, both hardware and software, may be traded remotely.

As part of the Market for Computational Services project [14], a number of ICENI compatible Web services have been developed to support economic activity [5]. These include the Chargeable Grid Service, a Grid Payment Service which acts as an interface to the underlying financial reconciliation systems, and a Resource Usage Service. The key technology used to support the trading of services is an Agreement Protocol that allows peer-to-peer negotiation of prices and deliverable services.

6. Projects Using ICENI

The ICENI system is being developed to support several applied application communities through the UK e-Science program. We illustrate two examples of our activity in the rest of this section.

6.1 RealityGrid

RealityGrid [23] is an Engineering and Physical Sciences Research Council (EPSRC) project that is exploring the use of Grid technology to investigate the mesoscale structure of matter. One application used in this project is LB3D, a parallel Lattice Boltzmann code written in Fortran 90 and C. ICENI is used to support LB3D, enabling a level of flexibility, Grid deployment and collaborative application usage unavailable with other systems. To aid the RealityGrid project, a complete application pipeline has been implemented, allowing the deployment, scheduling and visualisation of LB3D applications through ICENI.

In order to deploy LB3D on an ICENI managed Grid, the LB3D application is wrapped up in an ICENI 'binary component'. This component wrapper defines application meta-data to support execution of the binary within the ICENI infrastructure. Depending on the resource that has been selected, the LB3D binary component will either be run interactively or submitted to a batch system such as Sun Grid Engine [24]. It is important to note that though the ICENI component model cannot handle parallel components, the code wrapped in the binary component can be a parallel code.

Following the composition and scheduling of an application, ICENI facilitates the visualisation and steering of a running LB3D application. Much of the science enabled by LB3D concerns features (such as the gyroid phase) that are very difficult to discover automatically from the data; such features are only discernible through visualisation. Since ICENI exposes running components as services, collaborators can discover an executing application, and connect visualisation components at runtime without interfering or linking with the initial application.

Other activity within the RealityGrid project has involved development of a steering library to allow LB3D computations to be dynamically steered at runtime [4]. This library provides instrumentation and hooks into LB3D (and many other RealityGrid applications), and provides the link from the infrastructure to the application. As an ICENI component, the steering library is published

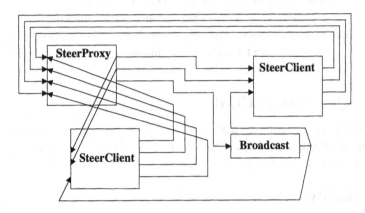

Figure 3.   Component composition showing a steering proxy used to support multiple steering clients.

as a service, and may be discovered and invoked by anyone with the correct access privileges. By utilising ICENI, multiple clients can invoke the steering library by connecting to the ports of its component, and their commands are passed in a coherent way to the application. This provides added value over using the steering library alone, which can only support a single client without interference. Figure 3 shows a component composition consisting of two steering client components, that may be running at different physical locations, connected to a steering proxy, facilitating collaborative steering.

Moreover, a facility is provided to stream the video of the visualisation through the Access Grid. The Access Grid nodes are started as ICENI components. The control of the different nodes is done centrally through the Access Grid Controller, another ICENI component. Any modifications to the Access Grid session from one node are automatically propagated to all other nodes. When defining the Access Grid application, it is possible to give each display node permission to control the Access Grid session (simply by making a specific port connection).

The ICENI Access Grid components also provide encryption of the data through a video and an audio key. These keys can be generated from any control component, they are then sent to all the display nodes that are part of the session. Any third-party user connected to the AG room will then be unable to connect to the ICENI session.

6.2   GENIE

The Grid ENabled Integrated Earth system model (GENIE) project [8] is funded by the Natural Environmental Research Council (NERC) and aims to deliver both a flexible Grid-based architecture, which will provide substantial long-term benefits to the Earth System modelling community; and also new scientific understanding of the global environment from versions of the Earth system model (ESM) developed and applied in the project.

The scientific focus of GENIE is on long-term changes to the Earth's climate, particularly the (geologically) recent ice-age cycles and the future response to the Earth system to human activities, including global warming. A realistic ESM for this purpose must include models of the atmosphere, ocean, sea-ice, marine sediments, land surface, vegetation and soil, ice sheets; and the energy, biogeochemical and hydrological cycling with and between components.

Consequently the e-Science objectives of the project are to develop, integrate and deploy a Grid-based system to flexibly couple together state-of-the-art components to form a unified ESM, execute them on heterogenous distributed computing resources, and share the resultant data in a distributed way, while still maintaining high-level open access to the system to support a virtual organisation of Earth System modellers.

ICENI has been used to meet some of these objectives [10]. In particular, we have been able to wrap a prototype version of GENIE, comprising of a 3-dimensional (frictional geostrophic) ocean model coupled to a 2-dimensional (energy-moisture balance) atmosphere model, as an ICENI Binary Component in order to perform parameter sweep experiments (see Figure 4). A Setup Component initialises the experiment, creating the necessary input files for the GENIE binary executable using parameters chosen by the user during the composition phase. This passes data to a Broadcast Component which delegates the data to multiple Binary Components (only 3 are shown in the figure), each of which execute the GENIE model. As each simulation finishes, the Binary Component passes the resultant data to a Funnel Component, which passes it to an Archive Component that handles the archiving of the resultant data.

In order for this application to run efficiently the ICENI framework provides a number of launching mechanisms that take the presence of Distributed Resource Managers (DRMs) into account during the scheduling and launching phases of application deployment. This allows a large number of jobs to be submitted to a single computational resource hosting a high-throughput system such as Condor [26]or the Sun Grid Engine. Moreover, the componentised nature of ICENI applications (see Section 3) allows us to submit a GENIE parameter sweep experiment across multiple computational Grid resources, and consequently we may combine and use several high-throughput systems in a single experiment.

As a result of our efforts, the environmental scientists involved in the GENIE project are able to commit their own individually administered computational resources to form a Grid and perform parameter sweep experiments consisting of approximately 1000 individual GENIE simulations per experiment. Thus al-

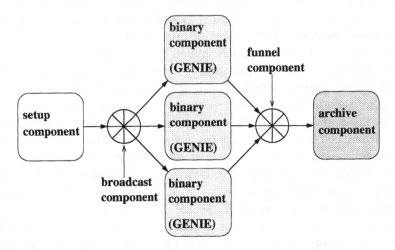

Figure 4.   The GENIE parameter sweep experiment as a component-based application. Arrows describe the direction of control and data flow between components.

lowing them to carry out scientific investigations about the natural environment in a relatively novel and significant way (e.g. [15]).

6.3   Further Improvements

Both the RealityGrid and GENIE projects can benefit from the performance and reservation mechanisms provided in ICENI. Both of these applications are based on the execution of a binary component, and have substantial computational requirements. As such the choice of the resource used to execute the component is critical. Performance data will help to select the best resource and advanced reservation will ensure the component has exclusive access to the resource.

Performance and reservation mechanisms provide the ability for each Grid user to optimise their use of the Grid. However, this approach on its own need not lead to an overall optimal use of the Grid. It may lead to conflict over resources, in which all users attempt to deploy their work to the same set of "best" resources. Alternatively this may lead to a situation in which low priority users unnecessarily take the "best" resources when their jobs arrive before those of higher priority users.

By adding an economic model to the Grid, the user is forced to balance their cost–benefit requirements. A user will wish to use the cheapest resource for which choosing this resource will not adversely affect their requirements. Likewise a low priority user is unlikely to have the funds to affect the requirements of a high priority user.

For projects such as RealityGrid which have high priority (interactive) users requiring real time performance, economics provide the mechanism to allow these users to obtain time on high performance (price) resources. Conversely in the case of GENIE where the results are not required to be interactive, "money" can be used in the purchasing of high throughput resources.

7. Related Work

In this section, we present some related works that offer component models intended for high performance Grid computing, similar to that of ICENI. The ICENI middleware system offers other functionalities alongside the component model; there are higher level services such as schedulers, resource reservation systems built upon a loosely coupled service oriented architecture, yet it is instructive to consider alternative component systems.

GridCCM [22] is a parallel extension to the CORBA Component Model (CCM): it defines a parallel component as a collection of identical sequential components that execute all or some parts of their services in parallel. Hence, parallel codes can be embedded into a parallel CORBA component with few modifications to parallel codes. Besides, GridCCM targets to extend CCM but not to modify the model so as to define a *portable* extension that can be added to any implementation of CCM. GridCCM supports parallel communication flows between parallel components, which may involve data redistributions.

The deployment model of GridCCM is based on the standard CCM deployment model. However some parts are being specialised for Grids, such as the interaction with a given Grid resource allocation environment, Globus [13] for example. It should be noted that the GridCCM implementation supports parallel components - a feature which is currently unavailable within ICENI.

ProActive is a middleware for object oriented parallel, mobile, and distributed computing. ProActive implements a hierarchical and dynamic component model, named Fractal [18], with the aim to master the complexity of composability, deployment, re-usability, and efficiency of Grid applications [3]. This defines a new concept of hierarchical components that, recursively, can be parallel, made of several activities, and distributed. These components communicate using typed one-to-one or collective method invocations. At runtime, newly deployed or discovered components can be incorporated in the application, as component bindings and component inclusions can be modified. Component mappings can also be modified due to the mobility feature of active objects implementing components. ProActive provides a critical feature that ICENI components and GridCCM components lack, the ability to seamlessly migrate components during execution.

The Common Component Architecture (CCA) [2] presents a unified component based model for high performance computing. There are a number of distinct implementations of the CCA, each featuring distinct functionality, such as distributed or parallel underlying frameworks. The CCA is designed around a spatial workflow model in which components exist concurrently, though current implementations of the CCA can be mapped onto temporally ordered Web service workflows. Such a model is similar to the ICENI component model, though the semantics are marginally different (standard ICENI component connections are one-to-one, with specific exceptions made for utility components such as broadcasts or splitters, while CCA components allow one-to-many connections).

There is a clear synergy between these four different component systems: ProActive, GridCCM, CCA and ICENI. They all have unique features which might be interesting to integrate in the other systems. We are currently examining the state of interoperability between these component systems and the ICENI component architecture, and exploring possible opportunities to layer the ICENI higher level services on a component framework from one of these related technologies.

8.   Conclusion

Within this chapter, we have presented recent work on the ICENI Grid middleware. We believe this represents one of the first attempts to provide an open and extensible service based Grid architecture that is populated with services that deliver usable functionality to e-researchers. These services range from simple job submission to the more advanced performance driven scheduling and reservation of complex multi-stage workflows. Sophisticated dynamic interactivity is enabled via the component based design model, which allows users to collaborate, steer, visualise and interactively modify their application workflows.

While work is still ongoing, we believe we have a framework that enables the non-trivial issues of building real usable Grid infrastructures, and have had demonstrable success in delivering added value to application scientists in multiple case studies.

References

[1] A. Afzal, J. Darlington, S. McGough, S. Newhouse, and L. Young, Workflow Enactment in ICENI. In *Proceedings of the Third UK e-Science All Hands Meeting*, Nottingham, UK, September 2004. http://www.lesc.imperial.ac.uk/iceni/pdf/ahm2004_WorkflowEnactment.pdf

[2] R. Armstrong, D. Gannon, A. Geist, K. Keahey, S. Kohn, L. McInnes, S. Parker, and B. Smolinski. Toward a Common Component Architecture for High-Performance Scientific Computing. In *Proceedings of the Eighth IEEE International Symposium on High Performance Distributed Computing, HPDC'99*, page 13, August 1999.

[3] F. Baude, D. Caromel, and M. Morel. From Distributed Objects to Hierarchical Grid Components. In *Proceedings of the International Symposium on Distributed Objects and Applications, DOA*, LNCS 2888:1226–1242, Springer, November 2003.

[4] J. M. Brooke, P. V. Coveney, J. Harting, S. Jha, S. M. Pickles, R. L. Pinning, and A. R. Porter. Computational Steering in RealityGrid. In *Proceedings of the Second UK e-Science All Hands Meeting*, Nottingham, UK, September 2003. http://www.nesc.ac.uk/events/ahm2003/AHMCD/pdf/179.pdf

[5] J. Cohen, W. Lee, A. Mayer, and S. Newhouse. Making the Grid Pay - Economic Web Services. In *Proceedings of the Workshop on Building Service Based Grids, GGF 11*, June 2004. http://www.doc.ic.ac.uk/ sjn5/GGF/GGF11/BGBS-Mayer.pdf

[6] J. Darlington, S. McGough, S. Newhouse, and L. Young. Performance Architecture within ICENI. In *Proceedings of the Third UK e-Science All Hands Meeting*, Nottingham, UK, September 2004. http://www.lesc.imperial.ac.uk/iceni/pdf/ahm2004_PerformanceArchitecture.pdf

[7] N. Furmento, J. Hau, W. Lee, S. Newhouse, and J. Darlington. Implementations of a Service-Oriented Architecture on top of Jini, JXTA and OGSI. In *Proceedings of the Second Across Grids Conference*, Nicosia, Cyprus, January 2004. http://grid.ucy.ac.cy/axgrids04/AxGrids/papers/E00-393596327.pdf

[8] The Grid ENabled Integrated Earth system model project. http://www.genie.ac.uk/

[9] Grid Resource Allocation Agreement Protocol. https://forge.gridforum.org/projects/graap-wg

[10] M.Y. Gulamali, T.M. Lenton, A. Yool *et al.* GENIE: Delivering e-Science to the environmental scientist. In *Proceedings of the Second UK e-Science All Hands Meeting*, Nottingham, UK, September 2003. http://www.nesc.ac.uk/events/ahm2003/AHMCD/pdf/026.pdf

[11] Jini Network Technology. http://wwws.sun.com/software/jini/

[12] The JXTA Project. http://www.jxta.org/

[13] S. Lacour, C. Pérez, and T. Priol. Deploying CORBA Components on a Computational Grid: General Principles and Early Experiments Using the Globus Toolkit. In *Proceedings of the Second International Working Conference on Component Deployment, CD 2004*, LNCS, 3083, Springer, May 2004.

[14] London e-Science Centre. A Market for Computational Services. http://www.lesc.ic.ac.uk/markets/

[15] R. J. Marsh *et al.* Bistability of the Thermohaline Circulation Identified through Comprehensive 2-Parameter Sweeps of an Efficient Climate Model. *Climate Dynamics*, Accepted, 2004.

[16] A. Mayer, S. McGough, N. Furmento, W. Lee, S. Newhouse and J. Darlington, ICENI Dataflow and Workflow: Composition and Scheduling in Space and Time. In *Proceedings of the Second UK e-Science All Hands Meeting*, Nottingham, UK, September 2003. http://www.nesc.ac.uk/events/ahm2003/AHMCD/pdf/132.pdf

[17] A. Mayer, S. McGough, M. Gulamali, L. Young, J. Stanton, S. Newhouse, and J. Darlington. Meaning and Behaviour in Grid Oriented Components. In *Proceedings of the Third International Workshop on Grid Computing, Grid 2002*, LNCS, 2536, Springer, November 2002.

[18] Object Web Consortium. Fractal. http://fractal.objectweb.org/

[19] Open Grid Services Architecture. https://forge.gridforum.org/projects/ogsa-wg

[20] Open Grid Services Infrastructure (OGSI) v1.0. http://forge.gridforum.org/projects/ggf-editor/document/draft-ogsi-service1/en/1

[21] OpenLDAP Project. http://www.openldap.org

[22] C. Pérez, T. Priol, and A. Ribes. A Parallel CORBA Component Model for Numerical Code Coupling. *The International Journal of High Performance Computing Applications*, 17(4):417–429, 2003.

[23] RealityGrid Project. http://www.realitygrid.org/

[24] Sun Grid Engine (SGE). http://wwws.sun.com/software/gridware/

[25] Sun   Microsystems   Inc.   Java   Web   Start   Technology. http://java.sun.com/products/javawebstart/

[26] D. Thain, *et al.* Condor and the Grid. In F. Berman, A. J. G. Hey, and G. Fox, editors, *Grid Computing: Making The Global Infrastructure a Reality*, John Wiley, 2003.

[27] W3C, Web Services. http://www.w3.org/TR/ws-arch/

[28] L. Young, S. McGough, S. Newhouse, and J. Darlington.   Scheduling Architecture and Algorithms within ICENI.   In *Proceedings of the Second UK e-Science All Hands Meeting*, Nottingham, UK, September 2003. http://www.nesc.ac.uk/events/ahm2003/AHMCD/pdf/005.pdf

AN ARCHITECTURE FOR A PORTABLE GRID-ENABLED ENGINE

Bruce Long and Vladimir Getov
Harrow School of Computer Science
University of Westminster
Watford Rd, Northwick Park
Harrow, London, U.K.
B.D.Long@westminster.ac.uk
V.S.Getov@westminster.ac.uk

Abstract   A common vision for the Grid is that it would enable heterogeneous, spatially distributed, non-trusted computers to operate as a single system for users, compilers, applications and services. However, currently available Grid solutions, while providing for such computers to cooperate, are complex and do not support the single system image principles. In this chapter, we describe a simple architecture for a portable Grid engine which produces and uses both interpreted and binary objects constructed around text-based mathematical descriptions of those objects and their classes. A first prototype engine has been implemented that achieves some functionality and a fully functional model is under development.

Keywords: Grid engine, Grid platform, OGSA, virtual organization, global namespace, lightweight platform

1. Introduction

The ability to access information on the Web from various types of devices anywhere in the world has been capturing the minds of individuals, developers, scientists, and those in business. Many particular Web systems already allow such access in various application domains. The next natural step in exploiting the Internet's huge opportunities is the development and utilization of the Grid. The emphasis in a Grid environment is the access to different kinds of remote resources in addition to simply information, dramatically increasing the variety of applications offered to the end user. While a rich collection of protocols and APIs have already been developed to facilitate Grid programming, there is no clearly defined *Grid platform* yet. As more complex solutions are requested by users, more tools and patches will be added, and, in the end, systems will be very complex webs of overlapped and obsolete tools, APIs, and protocols requiring engineers with very complex skill-sets. They will not grow well.

In order for a platform to offer available-from-anywhere Grid services on any information device and yet grow well, the functionality that provides such services must be factored completely out of the application software and into the platform. In particular, it would be useful to have a platform on which information could be stored and on which programs could be written, built and executed, but where the platform exists and is accessible on a system of loosely connected devices rather than on a single tightly controlled computer. There are currently several projects working toward this goal. For example, the draft report for an Open Grid Services Architecture (OGSA) platform [8] specifies that Grid services be built and managed according to the Open Grid Services Infrastructure (OGSI) [6] specification. It describes a number of platform interfaces which applications may use to find and manage data, policies, security information, etc. Lastly, it is also proposed in the OGSA platform draft specification that XML Schema Language (XSL) [7] models of common resources be used to standardize the use and management of common resources.

The draft OGSA platform specification does not express how, on an arbitrary host and in a heterogeneous environment, tasks can be run, binaries that access local resources can be created and executed, and a user interface can be accessed, perhaps on a non-local host. Part of this functionality is, however, implemented in the Globus Toolkit [9, 14]. A natural next step would be to develop the architecture for a portable Grid-enabled engine conforming to the following basic requirements:

- Portability

- Performance

- Small memory image

- Real-time support

This chapter outlines a solution in which an abstract modeling language is used as an intermediate language to respond to changes in an environment by choosing from among a variety of solutions such as building a specialized binary to take advantage of new hardware or working with a different user interface paradigm. The described solution provides a single system image (SSI) [3] for both service and application binaries by having languages compile to the intermediate language. The rest of this chapter is organised as follows. Section 2 provides an introduction into different types of environment bindings and the modeling language – Quanta. Section 3 discusses the model of the generic Grid engine, while Section 4 describes its architecture. The implementation methodology is covered in Section 5, followed by Section 6 on related work and the conclusions.

2. Environment Binding

Since many of the resources and services upon which an application or service relies during execution vary in format from environment to environment, a generic platform must be able to identify any environment-specific functionality and make an appropriate substitution of identical functionality at build time, load time, or even during execution. To accommodate such specialization, we mention three types of binding common to most environments.

- Binding to local resources. High-speed local resources such as graphics hardware, FireWire ports, or cards on the local bus can be utilized more efficiently if the binding to them is not through a Grid service on the local host.

- Binding to build and run binaries or perform calculations. The binaries used to perform tasks may vary over environments as well as over the method of executing the task. On a low level, binaries must be built per environment to bind to the local operating system or processor type. For example, an operation to process a block of data may be implemented as a loop on one machine but as a parallel operation on another.

- Binding to a user interface platform. Applications or services that must interface with users on a heterogeneous system must be bound to a local user interface platform such as the Windows GUI, Java Swing, or a Web-based user interface. If the user is not at the local machine, as is common with distributed gaming, programs may be bound to a user interface proxy.

2.1 Binding via Identity Substitution

The method being described to bind programs to an environment is to represent situations that may vary over environments and, at the appropriate time, substitute an environment-specific situation that provides equivalent functionality. In the simplest case, such a situation may be of a call to a function or an assignment operation. More complex binding operations, such as binding to custom processing hardware or specializing the creation of a binary, may require substitutions through sequences, conditionals, or repetitions. With the addition of nested substitutions, the creation of complex compositions of services, local resources, and user interface components may be automated.

2.2   The Quanta Language

Quanta is a lightweight core language [17] for describing objects and classes, and systems of objects and classes. Based on this description, an engine can then be used to make inferences about the objects and classes in order to instantiate objects or to query and manipulate them. By extending the engine's namespace to include other namespaces or outside objects, Quanta and the engine can be used to manipulate and query such systems as easily as internal objects. A variety of methods for accomplishing a task can be represented, thus allowing the engine to determine an optimal method. Lastly, a Quanta engine can be made to translate sequences, conditionals, and repetitions into code, providing the ability to produce a C file or a binary specialized for a particular environment. It can be used as an intermediate language by compiling, for example, C++ or Java byte code to Quanta.

Quanta models classes, objects, and systems of objects in such a way that the effects of using a modeled system can be inferred. Objects classified can be digital or analog, physical or abstract. A Quanta engine can use the models to infer what use of resources on the local system or on a network can be substituted to meet given binding requirements. The engine then makes the required substitutions. Because the engine can make substitutions involving the von Neumann structures of sequences, conditionals, and repetitions of information manipulations, entire algorithms can be reworked as needed to take advantage of new resources and accomplish a task or instantiate an object. The following are some of the major components of the Quanta language:

- Numeric, string, and Boolean literals

- Names / functions

- Identity assertions

- Informatic membership, union, difference, intersection and complement

- Block, repetition, and conditional constructions

- A connection to the local system, e.g., ability to make certain system calls.

Using identity assertions, any construction of the above components can be hierarchically mapped to any other one. For example, a function can be mapped to an algorithm or to a system call, while a repetition or sequence can be mapped to the results of running a parallel processor using a given algorithm.

3. A Generic Integrated Grid Engine Model

The generic Grid engine model being described has three aspects: namespace management, the execution environment, and binary-to-environment binding. While the engine is executing on a local host platform, it is running in the context of one or more virtual organizations (VO) [11]. In addition to any VOs with which the platform engine associates due to membership or affiliations of the owner, a global VO will exist and operate in much the same way that peer-to-peer services such as Kazaa operate, yet without the purpose of mass file sharing.

3.1 Namespace Management

The platform engine running on its local host will maintain a hierarchical namespace, as shown in Figure 1. The local namespace will include references and Quanta descriptions of such items as the following:

- Local hardware available

- Local file systems and objects/services

- References to local users, their local settings and objects

- Local settings

In addition, each VO of which the system is a member will provide resources to the namespace. In particular, the global Grid VO will provide a global namespace managed by the peer-to-peer cooperation of all the systems in the VO. By storing files and other objects in their global Grid namespace folder, users and VOs will be able to access and manage their information globally, even if their own machines are currently down. Other entities referenced in the namespace are running processes and version information through Quanta descriptions.

3.2 The Execution Environment

A process begins executing when it is created in the namespace and its "executing" property is set to "true". Other properties express whether it is running on the local machine, a remote machine, or on multiple machines from one or more VOs.

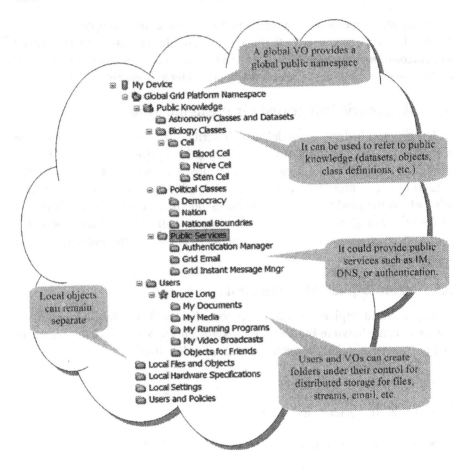

Figure 1. A device namespace

To support the widest variety of situations and preferences, there are three executions modes (see Figure 2) under this proposed generic engine model. In the "direct model" mode, the algorithms to be executed or the queries to be run are specified as Quanta models. The Quanta engine decides how best to execute them, whether to do so itself, build specialized binaries and execute them on remote machines, or execute pre-specified processes or Java code [12] on remote machines via Grid resouce allocation and management (GRAM) [13, 19]. Perhaps a particular task will be divided into parts and executed on different machines by all three methods. In "intermediate language" mode, programs written in traditional languages such as Fortran, C, C++, Perl, or .Net CLR can be compiled to Quanta as an intermediate language. The Quanta code can then

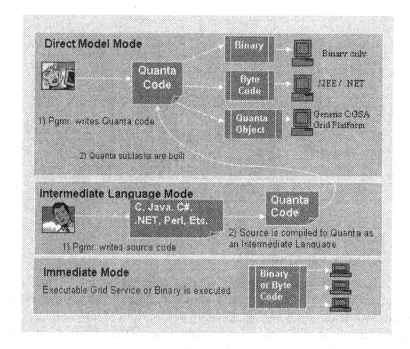

Figure 2. The three execution models

be executed in "direct model" mode. Lastly, in "immediate" mode, binaries and Java code built by other means can be executed directly using GRAM.

3.3 Binding Specialized Binaries

The heterogeneous nature of Grid computing requires generalizing over both hardware and user interface types. By using an intermediate language which can model hardware, user interface types, and applications, the portable Grid engine may build specialized binaries for each new machine and user interface types. As shown in Figure 3, there need not be a Grid user interface. Instead, each situation can be mapped to the location, available user interface platforms, and preferences of the user.

4. Architecture

The engine is built from a small number of simple components that achieve functionality by creating or utilizing objects described in various Quanta documents. In addition, Quanta documents describe users, policies, and services to be provided.

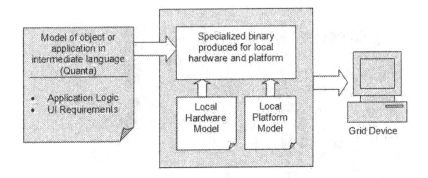

Figure 3.   Building specialized objects

4.1   Engine Components

The engine architecture is built around a generic platform engine, which interprets the descriptive documents, as shown in Figure 4. The platform engine uses the class descriptions and instantiation descriptions of resources contained in the various documents to organize manipulations of the resources available to it to accomplish the tasks given it. The platform adaptor interfaces the generic engine to the host system, the Grid services, user interface, and to the system Web server. While the generic engine remains the same regardless of the host platform or the user interface mechanism, ports of the system to various host platforms will involve modifying the platform adaptor and corresponding host environment documents to interface the generic engine to the host system, the Grid services, user interface, and a Web server. For example, a version running on a Microsoft Windows system may have an adaptor, which allows the engine to make win32 operating system calls. A corresponding document in the host environments documents would describe the relevant calls to the engine. In addition, a document describing how binaries can be created on a Windows host would be included.

The Web server component allows the engine to communicate with other engines as well as with remote users via Web pages. It also provides the interface for offering Grid services to other computers. For ports of the engine to other host platforms, it is likely that the Web-server component will have to undergo minor modifications to build and execute. Thus, a port to another system type involves modifying the engine adaptor, creating several host environment documents, making any necessary modification to the Web server code, and creating an installer program for that system type. A porter may include a transport mechanism other than http; however, a Quanta document describing the protocol should be included in the collection of host environment documents so that the engine can use it accordingly.

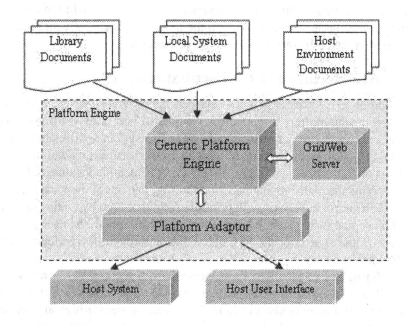

Figure 4. Engine architecture

Functions from the host system which are made available might include functions of the host operating system such as access to the file system, TCP/IP networking, I/O to hardware such as printers or graphics hardware, synchronization mechanisms, or control over processes and threads. The decision will be based upon the purposes of the porter. Whatever subsystems are made available, a corresponding Quanta document specifying the functionality to the engine should be added to the collection of host environment documents. For some devices, access to devices through Grid services may be sufficient, and thus some ports may not provide any access to the local host system.

Access to the host user interface functions may also be controlled by the porter; however, a sufficient subset of user interface components should be available if the engine is to implement such functionality as dialog boxes, text editing, etc. Some devices such as a small cell phone or a device imbedded in a simple sensor may not need or be capable of a full user interface. Likewise, for some applications, the user interface provided by the Web server may be sufficient. A Quanta document describing the user interface system should be included in the collection of host environment documents.

The above architecture achieves portability across types of host systems by abstracting the host systems and user interfaces. Performance and real-time support are achieved by providing for the engine to accomplish tasks through the creation of binaries that can be optimized. The system will have a small memory image because the engine and the adaptor are relatively simple, and, when needed, the Web-server module can be made small.

4.2   Architecture and the User Experience

In addition to the standard OGSA services, the generic platform engine running on a machine offers a number of services to both users and other such running engines if they are in a shared VO. The system will retain policies concerning which users are offered which specific services, as well as which machines or VOs can request them. When a user is identified at one of the connection points the platform engine is monitoring, perhaps at the local machine or by a Web interface, services will be offered and delivered by finding or building a binary that instantiates the requested application with the user interface required. Some aspects of the task may be done through interpretation rather than building a binary.

By offering and expecting certain services from peers, a VO or a union of VOs will be created with a SSI, which can carry out tasks or offer services independent of any particular machine. Thus, once a user starts an application, it may not be apparent which machine or machines are actually carrying out the computations or storing the information. Because messages from users are marked as originating from a user, not a machine, in the namespace, the users can switch machines in the middle of an application and have the user interface follow them and even adjust for heterogeneity.

The decentralized nature of such a process, together with the existence of a global VO, makes possible applications such as Grid-based DNS, public authentication, or decentralized instant messaging to run outside the context of a hosting organization. It also makes a high degree of cooperation possible; for example, while a cell phone may be able to broadcast video over TCP/IP to one or two viewers, if it were broadcast to a folder in the global VO (perhaps the folder represents a unified resource idenifier), the VO could ensure that the broadcast was replicated at strategic points and could serve the video at almost any scale.

4.3   Examples of Models for Various Host Environments

The models that populate the system's namespace are held and communicated in Quanta documents. Based on their use and lifetime, four categories of documents can be identified. Below are examples of documents that might exist in a finished product; they are in no way prescriptive and may be altered considerably in the final platform engine design.

Library documents are static and available on all engine hosts. They include models of mathematical classes, identities and functions. Other library documents may describe typical computer hardware, data structures and algorithms. Objects such as queues, strings, and streams, as well as memory, network nodes, and common CPUs may be described. Also, documents may describe Grid concepts such as the OGSA protocols, OGSI and other API's and utilities. For example, a Quanta document might map names and functionality to LDAP [15] or UDDI [14] documents to facilitate the use of such services. Those document maps would be built on a lower Quanta document mapping the names implied by XML [20], XSL and intermediates such as SOAP [2].

Local system documents give the engine enough information to identify local users, use local hardware, understand policies, offer services, and log on to and participate in VOs, including the global VO. Such documents might describe local hardware, local users, services to be offered, and to whom services should be offered.

Host environment documents may describe particular types of hosting environments such as Microsoft Windows, Java, or SUN environments. Documents detailing how to compile and build binaries or C programs are an example, as well as one detailing the execution environment; how to run and interact with processes. Also, any local user interface types can be detailed.

Lastly, a number of *Quanta and WSDL [4] documents* will be exchanged when an engine connects to any VOs, including the global VO. Such documents might include models expressing network topology, authentication information, collective namespace negotiation, or the locations of GIIS [5] servers.

5. Implementation Methodology

There are two aspects of the use of the Grid engine which must be discussed: individual engines running on single machines, and the cooperation of the individual engines to produce a distributed platform. First we shall discuss the implementation of a single engine.

5.1 Implementing an Individual Engine

On a high level, there are three major aspects of the engine that must be implemented. First, there must be a representation of the namespace. Second, there must be a method of processing the items named. Lastly, there must be a mechanism providing access to items in the namespace from the outside world to communicate with people, other computers, databases, sensors, etc. There is an obvious analogy with the system architecture which consists of computer memory (the namespace), a CPU, and I/O ports.

5.1.1 The Namespace. The namespace of our current implementation of the Quanta engine is stored with a C++ class called *infon*. The properties of infons are formally described in [18], though that document is currently a draft. Briefly, infons can have identity and containment relations to other infons. There are operators which join infons to each other in various ways to produce new infons. Operations such as intersection, union, and difference are defined. Infons can also be associated with a name. These properties can be represented in a programming language by giving infons a linked-list of their identity relations, is-a/has-a relations and the negation of those relations. Literal infons have a pointer to raw memory or some other sequence of bytes. If the infon contains subparts such as with a C struct or a union, these are spelled out as idenitites and containments. For example, a literal infon representing an email address would contain a user-name infon and a domain-name infon. A non-literal infon for an email address would contain the same parts, but they might be mapped to (identical to) a record in a database or to a field in an HTTP form, for example. The syntax of Quanta maps directly to the structure of an infon. That is, when Quanta is loaded it is an infon, and when an infon is serialized it is in Quanta. The low level properties of infons can be used to create complex layers of new infons.

In our implementation, an infon called World forms the top-level container for the namespace of the engine. When the engine loads, it populates World by loading a top-level Quanta document that includes the other Quanta documents such as ones describing local resources, users, etc. World normally undergoes constant modification as items within it change state or new items come into or leave the namespace.

5.1.2   Processing.   In the current implementation of the engine, the processing of the infon World is done with an object of a class called *agent*. The agent creates World and loads it from the top-level Quanta documents. As the agent loads World, it identifies computing tasks which should be done. For example, it may be that monitoring port 80 is required or that a particular algorithm must be executed. These tasks are submitted to an agenda to be executed by other threads. When a task is popped off of the agenda, the engine searches World for actions within its capability that would be identical to the task as it is described. Often it will find a number of different methods which would be suitable. Each may have unique time/space requirements. One of the solutions is chosen and executed. The results of the execution are often imbedded back into World causing new tasks to be added to the agenda for execution. The details of this processing are given in [17].

5.1.3   Input/Output.   Perhaps the simplest part of the engine is the method of connecting it to the I/O system of the host computer. The agenda can substitute specific (as determined by the implementation) Quanta functions with a call to a C function or an operating system call. For example, when a task of producing a "beep" sound is popped from the agenda the system will find that one of the possible ways of producing a beep will be to call the host system's operating system. This method of tying Quanta functions or objects to C or system functions or objects is used to receive input as well.

5.2   Implementing Cooperation Among Groups of Engines

To create a VO, it is necessary that the individual engines in the group cooperate. A naive statement of the problem might be:

1 There exists a collection of goals to be fulfilled;

2 There exist resources for fulfilling those goals;

3 There exist methods for locating, accessing and utilizing the resources.

In practice, however, goals are often contradictory, available resources are finite, and methods for accessing and utilizing resources do not always work.

One method of implementing cooperation among engines is for each engine to interact with other engines in the same way that they interact with, for example, Web or Grid services. That is, to use a WSDL or Quanta description of the service to understand what it does, call the service, verify the results and the trustworthiness of the service provider, then accept the results. There are two drawbacks to using this as the only available method of cooperation. First, this method places limits on how closly engines can cooperate on a task. Second, it does not truly conform to the goal of having a program running on a distributed platform. Rather, the program is running on whichever machine is hosting the engine which processed the particular service request, and it farms out parts of the task to other machines. There are two other implementation levels of cooperation which extend the flexibility of the system and which deal with contradiction and uncertainty in a manner that may be more robust in an peer-to-peer environment.

The second level of cooperation is to allow engines to share a single namespace and a single task-agenda in some way. They could use a shared data structure or a virtually synchronized data-structure. The current plan is to have engines negotiate what to share when they first contact each other, then to maintain the share as needed. Likewise, cooperating engines might share a single task agenda such that it is unknown which machine will be running a particular task until one of them pops it off of the agenda. Such a shared agenda could also be used to allot some tasks to more than one machine and compare the results to test for trustworthiness.

A third level of cooperation is possible that provides for processes to execute in a truly distributed fashion. With shared namespaces and shared agendas, processes are running on multiple machines. It would be useful to have these no-particular-machine processes be in charge of such tasks as maintaining the system and providing user services. This could be done with some redundancy such that rogue nodes, failure, and uncertainity could be mitigated without relying on a "manager" machine.

Furthermore, by moving the logic of cooperation, authentication, and security into the engine, not only is the architecture simplified, but the system can begin to handle novel problems such as distributed denial of service attacks by devising a solution using the same tools it uses to devise solutions to user problems.

5.3   Examples

The following two examples are selected to illustrate our implementation methodology.

5.3.1   Example 1.
Suppose a user queries a PocketPC, which is running a Grid engine, for an estimate of the number of stars in our galaxy. The engine will first search its own namespace. Next, it will query a Quanta-Grid supernode by negotiating a partial namespace merge with the supernode and running the query on it. This may call data stores on OGSA nodes or use an XML-based Web service. The result will be that the supernode returns some Quanta code to the PocketPC's engine. The code may be an actual number, but more likely it will be an expression that could be as complex as the following meaning: "Send the following query string to the computer at 221.45.65.2 using http POST, then use the following algorithm to decrypt the results returned. Use SSH to contact the computer at SciData.org and send it the results of the previous query. You will receive a number (via the following special protocol) that is 1/2 of the result you seek." The local engine would merely evaluate the expression it receives and in so doing, carry out the necessary actions.

5.3.2   Example 2.
Suppose an end user engages a local Quanta engine to carry out a very complex simulation. By examining the number of items in the collections it would have to instantiate, and by looking at the functions it would have to call, suppose the engine calculates that it does not have the resources needed to finish the task in a reasonable amount of time. Suppose then that the local engine contacts a supernode and merges its namespace with the part of user's personal namespace which is the simulation. As it happens, that object is a running program that can be divided into many threads and which will solve the problem. The supernode checks quotas and allocates an optimum number of computing resources on machines that the user is authorized on to instantiate the object. Each machine uses an engine to optimize the code it runs for its own architecture. The supernode may relinquish the monitoring of the task back over to the local engine, or it may allocate it to yet another machine or supernode. However it happens, the results will be collected and presented to the local machine in some suitable format as requested. The user

can then ask that they be printed in home and work offices as a graph and have the corresponding spreadsheet printout faxed to a colleague from the address book. Later, the graph of the data could be inserted into a document stored in the namespace. All of the above interactions involve only the manipulations of a large shared Quanta namespace by the local device, the Quanta supernodes, and the slave machines that provide Web or Grid services.

5.4 Current Status

A prototype engine was constructed which implemented the major functionality of the Quanta engine. However, this engine suffered from a number of problems. Among others, it could not work with streaming data and it scaled very poorly. The version of Quanta that it processed had a purely hierarchical namespace which was not graceful under namespace collisions – a clear disadvantage at the global scale. The engine was redesigned to account for lessons learned the first time, and the Quanta language has been simplified by adhering closer to the theory in [18]. In addition to making the core of the Quanta language smaller, the simplification makes it handle namespace collisions more gracefully. When the new engine is completed, formal experiments will be designed to demonstrate its functionality.

6. Related Work

There are a number of related projects in development which offer similar or complementary functionality to Grid engines as proposed here. A similar vision based on a component-oriented approach to building a generic Grid services platform has been suggested in [21]. In this work a core collection of components would provide a foundation of functionality to support Grid applications. Higher level components would be built upon this core. A knowledge engine is used in order to track available resources, and a component manager manipulates the resources to create new components. This approach has much in common with the approach presented here; however, there are differences. The primary difference is that our model utilizes a generic information processing engine (the Quanta engine) which can create objects and perform actions based upon Quanta descriptions in text files. This allows the knowledge engine, the component and service managers, and the core components to be collapsed into the Quanta engine plus several Quanta documents.

Globus (version 3) [14] is based upon OGSA Grid services and is seen as complementary to the current project. Nimrod/G [1] is a Grid-enabled tool for performing parametised simulations which is built upon Globus.

7.   Conclusion

As presented in this chapter, the problem of performing an action in a hetero-geneous environment could be solved by identifying collections of situations which, if they obtained, would be identical with regard to a desired outcome. Next steps involve choosing and executing by substitution one situation from that collection which is compatible with the current host environment. This "situation-substitution" scenario can be completed by the programmer with conditionals, or at build-time, load-time, or runtime. Since the logic of substi-tuting such "identicals" does not vary whatever the problem or time, rather than creating a special tool or application for each situation, the Quanta language can be used to specify which situation-substitutions are sufficient to instantiate a solution to a particular task. The Quanta engine, when coded to be a Grid engine, can be made to find optimal substitutions and make them, whether at code-time, runtime, or at some time in-between, in the context of the global Grid. Such a system would facilitate the use of new and legacy code in many languages, provide an extensible, global namespace, and change the perspective for Grid-based applications and services from that of "running on a local host while accessing distributed resources" to "running on the Grid".

References

[1]   D. Abramson, J. Giddy, and L. Kotler. High Performance Parametric Modeling with Nim-rod/G: Killer Application for the Global Grid?, *Proc. of IPDPS*, pp 520–528, Cancun, Mexico, May 2000.

[2]   D. Box, et al. Simple Object Access Protocol (SOAP), *W3C*, http://www.w3.org/TR/SOAP/, May 8, 2000.

[3]   R. Buyya, T. Cortes, and H. Jin, Single System Image (SSI), *The International Journal of High Performance Computing Applications*, 15(2):124–135. 2001.

[4]   E. Christensen, F. Curbera, G. Meredith, S. Weerawarana. Web Services Description Lan-guage (WSDL) 1.1, *W3C Note*, http://www.w3.org/TR/2001/NOTE-wsdl-20010315, 15 March 2001.

[5]   K. Czajkowski, S. Fitzgerald, I. Foster, and C. Kesselman. Grid Information Services for Distributed Resource Sharing, *Proc. 10th IEEE International Symposium HPDC-10*, IEEE Press, http://www.globus.org/research/papers/MDS-HPDC.pdf, pp. 181–194, 2001.

[6]   K. Czajkowski, I. Foster, J. Frey, S. Graham, C. Kesselman, D. Snelling, S. Tuecke, and P. Vanderbilt. Open Grid Services Infrastructure (OGSI), Open Grid Service Infrastructure WG, Global Grid Forum, http://www.ggf.org/ogsi-wg/drafts/draft-ggf-ogsi-gridservice-26.2003-03-13.pdf, March 13, 2003.

[7]   D. Fallside. XML Schema Part 0: Primer, *W3C*, http://www.w3.org/TR/xmlschema-0 May 2001.

[8]   I. Foster, D. Gannon. The Open Grid Services Architecture Platform, *GGF-WG OGSI*, http://www.ggf.org/Meetings/ggf7/drafts/draft-ggf-ogsa-platform-2.pdf, February 16, 2003.

[9] I. Foster and C. Kesselman. The Globus Toolkit, In *The Grid: Blueprint for a New Computing Infrastructure*, I. Foster and C. Kesselman (Eds.), Morgan Kaufmann Publishers, San Francisco, California, pp. 259–278, 1999.

[10] I. Foster, C. Kesselman, J.M. Nick, and S. Tuecke. The Physiology of the Grid: An Open Grid Services Architecture for Distributed Systems Integration, *GGF-WG OGSI*, http://www.globus.org/research/papers/ogsa.pdf, June 22, 2002.

[11] I. Foster, C. Kesselman, and S. Tuecke. The Anatomy of the Grid: Enabling Scalable Virtual Organizations, *Intl J. Supercomputer Applications*, 15(3) http://www.globus.org/research/papers/anatomy.pdf, pp. 200–222, 2001.

[12] V. Getov, G. von Laszewski, M. Philippsen, I. Foster. Multi-Paradigm Communications in Java for Grid Computing, *Communications of the ACM*, 44(10):118–125, October 2001.

[13] Globus Project, GRAM: Grid Resource Allocation and Management, http://www.globus.org/about/events/US_tutorial/slides/Dev-06-ResourceManagement1.pdf, 2002.

[14] Globus Project. Status and Plans for Globus Toolkit 3.0, http://www.globus.org/toolkit/gt3-factsheet.html February 19, 2003.

[15] J. Hodges. An LDAP Roadmap & FAQ, http://www.kingsmountain.com/ldapRoadmap.shtml, December 6, 2001.

[16] K. Kennedy. Compilers, Languages, and Libraries, In *The Grid: Blueprint for a New Computing Infrastructure*, I. Foster and C. Kesselman (Eds.), Morgan Kaufmann Publishers, San Francisco, California, pp. 181–204, 1999.

[17] B. Long. Quanta: a Language for Modeling and Manipulating Information Structures, http://perun.hscs.wmin.ac.uk/pages/bruce/, December 2002.

[18] B. Long. The Structure of Information, http://arxiv.org/abs/cs.LO/0309004, September 2003.

[19] S. Martin. GT3 GRAM Overview, http://wwwunix.globus.org/ogsa/docs/alpha/gram/gt3_gram_overview.htm, January 8, 2003.

[20] L. Quin. Extensible Markup Language (XML), *W3C*, http://www.w3.org/XML/, February 26, 2003.

[21] J. Thiyagalingam, S. Isaiadis, and V. Getov. Towards Building a Generic Grid Services Platform, In: *Component Models and Systems for Grid Applications*, pp. 39–56, Springer, 2004.

[22] UDDI.org. Universal Description, Discovery and Integration, UDDI Technical White Paper, http://www.uddi.org/pubs/Iru_UDDI_Technical_White_Paper.pdf September 2000.

III

COMMUNICATION FRAMEWORKS

DYNAMIC ADAPTATION OF PARALLEL CODES: TOWARD SELF-ADAPTABLE COMPONENTS FOR THE GRID

Françoise André
IRISA/Université de Rennes 1
Rennes, France
Francoise.Andre@irisa.fr

Jérémy Buisson and Jean-Louis Pazat
IRISA/INSA
Rennes, France
Jeremy.Buisson@irisa.fr
Jean-Louis.Pazat@irisa.fr

Abstract   One of the challenges that come from the emergence of Grid architectures is to invent new programming techniques for these new platforms. As we explain in this chapter, we think that the architecture of the applications should reflect both the parallel and the distributed aspects of Grid architectures. It results in applications built as assemblies of parallel components. Since Grid architectures are known to be highly dynamic, using resources efficiently on such architectures is a challenging problem. Software must be able to react dynamically to the changes of the underlying execution environment. In order to help developers to create software for the Grid, we are investigating a model for the adaptation of parallel components. This chapter focuses on the adaptation mechanisms that are provided as a meta-level for components. We describe how a generic platform can help to develop efficient Grid software. First experimental results show the gain that can be expected from the use of such a platform.

Keywords: dynamic self-adaptation, parallelism, reflexive programming.

1.   Introduction

Research in Grid computing mainly focuses on the development of middleware and services allowing applications to use various distributed resources. Infrastructures and toolkits such as OGSA [3] and previously Globus [5] provide tools allowing naming and discovery of resources; they also provide the necessary services for applications to deal with the underlying heterogeneity of the Grid. Those projects provide the users with useful tools for deploying and running applications without explicitly dealing with the various batch queues, communication libraries and so on, installed on the local sites.

Although resource allocation and scheduling are taken into account, these tools give no help for applications to make efficient use of the available Grid resources at run time. Due to the dynamic nature of the Grid, it is also very hard to design an application that fits well with any configuration of the Grid. Moreover, constraints such as the number of available processors, their respective load level, available memory and network bandwidth are not static. The bandwidth between sites running some parts of the same application may vary during the execution time or some processors may be requested by other applications. For example, the CPU manager described in [8] may dynamically change the number of processors allocated to each application. We think that in many cases, application performance can be greatly improved when any part of an application can take into account varying resources eg. is able to adapt its behavior to "environmental changes".

Because Grid applications are also quite complex, many approaches now rely on service-oriented technologies such as OGSA [3] or on component-based approaches [11] such as CCM [9]. This helps building reusable software. Within these programming techniques, one of the main issue is the so called "separation of concerns" paradigm. Entities implementing distinct functionalities should be located in different modules, objects, services or components. In the remaining of this chapter, the term "component" stands for any of these kinds of pieces of application.

Our work focuses on the problem of adapting parallel codes encapsulated in components to varying constraints on resources. In this chapter, we show how to combine parallel programming and adaptation techniques in a unified framework. As a first step, this chapter focuses on the adaptation of the inner parallel code of a component. In Section 2, we describe parallel objects and parallel components. Section 3 is devoted to the presentation of adaptation techniques in the context of scarce resources, putting the emphasis on the description of the ACEEL framework. Section 4 presents the application model we consider. Section 5 explains how we transpose adaptation models to the parallel case to build a unified framework. The results of our experiments are given in Section 6. As a conclusion, we describe the main steps of our ongoing work.

2. Parallel Distributed Objects for the Grid

During the last decade, improvements have been made in terms of software reusability when object then component technologies appeared in many application fields other than high performance computing. Because Grid architectures are complex, heterogeneous and dynamic, they make the development of parallel applications more complex. Now, it becomes necessary for high performance applications to adopt technologies enforcing code reusability. Well-known component models such as CCM or Enterprise Java Beans should be a good basis but they were not designed with performance and parallelism in mind, so they have not been able to take into account High Performance Grid applications.

In order to improve these environments, several projects have studied how parallel code could be encapsulated within objects or components. Projects such as PARDIS [7] and PaCO++ [2] have focused on increasing performances of parallel distributed objects. They consider a parallel object as a set of identical sequential objects. The same definition also applies to GridCCM [10] within the component world.

Those projects allow to encapsulate SPMD code into so called "high performance CORBA objects/components". When a parallel object/component has to process a remote call, each process executes one part of the processing related to one part of the data set. The parallelism comes from the distribution of the parameters.

In order to get some performance out of high-speed networks, an enhanced request protocol has been defined among parallel objects/components: servers allow their clients to "see" their internal structure and distribution at run time. This allows parallel clients to send data directly from the source process to the target process: data do not need to be received/sent by a single master object. This multi-port communication mode allows to use the aggregated bandwidth, which can be higher than if only one centralized communication port was used.

These approaches have shown that it is now possible to use component-based techniques for programming high performance applications on the Grid without loosing performance. The next step is to be able to have components that are more flexible to allow the adaptation (not only the configuration) of parallel codes.

3. Dynamic Adaptation of Components

In the area of wireless computing and mobile environments where resources are a key issue, many techniques of dynamic adaptation have been developed:

from the observation of the environment, codes can adapt their behavior to fit the resource constraints. This adaptation can be achieved in many different ways ranging from a simple modification of some parameters to the complete exchange of the running code with a new one that is more suited to the environment.

The adaptation could be achieved by designing ad hoc applications that take into account the specificities of the targeted environment. For example, this was done for the Web applications access protocol on mobile networks by defining the WAP protocol [12]. A more general way to allow an application to evolve according to its environment is to provide mechanisms that permit dynamic self-adaptation by changing the behavior depending on the currently available resources. In many cases, this has been achieved by embedding the adaptation mechanism within the application code. For example, the AdOC compression algorithm [6] includes such a mechanism to change dynamically the compression level according to the available resources. However, it is desirable to separate the adaptation engine from the application code in order to make the code easier to maintain and to easily change or improve the adaptation policy. In this case, a framework that provides generic mechanisms for the adaptation process and for the definition of the adaptation rules is needed. This is the case for example of the ACEEL framework.

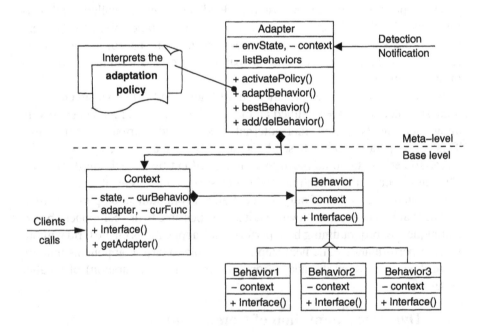

Figure 1.   Architecture of an ACEEL component

ACEEL [1] is a generic framework for self-adaptable components that allows the developer to focus on the implementations of the functionalities of his component and on its adaptation *policy*: it separates the adaptability aspect from the functional part of the component, as shown by the architecture of an ACEEL component in Figure 1. Based on the *Strategy* design pattern [4], the component offers a set of possible implementations, called *behaviors*. At any time, only one behavior is active: the one that processes the incoming requests. The generic *adapter* meta-object decides which of the available behaviors the best to use according to the environment is. To help the adapter object to decide, the component developer provides a policy as a set of event-based rules: each rule is a triggering change of the environment associated with a reaction, which might be either the activation of another behavior or the tuning of some parameters. When a change in the characteristics or availability of a resource happens, the monitoring engine notifies the adapter of the components that depend on this resource for their adaptation policy. The *context* holds the state of the component. Separating the state from the implementation makes it easier to replace dynamically the implementation of the component.

4. A Programming Model for Grid Applications

The applications we target in our project are "Grid applications"; they may be composed of several parallel codes, so in our model, an application is considered to be built as an assembly of components. Each component is deployed on a site that is a parallel machine such as a cluster. As a first step, we focus on the deployment and execution of one component and we do not investigate the relations between components. In our model, each component is both parallel and adaptable.

A component is parallel: this means that it is composed of a number of internal processes working together to execute a given service. These processes communicate between each other using a communication library such as PVM or MPI or through a distributed shared memory. Here, we do not require specifying how those processes are encapsulated within the component: this aspect relies on constraints of existing components platforms such as GridCCM.

A component is adaptable: the platform where components are deployed can monitor the resources of the deployment site and allow any component to react to any change in the state of the resources.

We think that it makes sense to allow one single component to adapt itself dynamically in most Grid environments for two main reasons. First, one characteristic of Grid architectures is that sites are administrated independently one from another and of the users of the Grid. It is thus possible that the site into which the component is deployed is modified while the component is running. Secondly, in a longer term, we can consider component migration as a special case of adaptation. Because the sites implied in the migration may have different characteristics, the adaptation of the component will be needed.

5.   Adaptation of Parallel Components

A parallel self-adaptable component is a component that is composed of several processes working together and that is able to change its behavior according to the changes of the environment.

5.1   Component Structure

The structure of parallel self-adaptable components includes an adaptation *policy*, a set of available implementations, called *behaviors*, and a set of *reaction steps*.

The purpose of the adaptation policy is to define when the adaptation mechanism should be triggered and what should be the associated reactions. It is mostly a set of event-based rules. Each rule associates a reaction to a specific event. Events represent any change in the state of the environment. For example, an adaptation policy can include the rule: "if the number of nodes is increased, spawn new processes and redistribute arrays". This rule shows both the trigger event and the associated reaction.

Behaviors are implementations of the component. Each behavior differs from the others in the way it uses resources and/or in the algorithm used. Each behavior of a component implements the whole interface of the component; they just use different ways. The active behavior is the one used to process the incoming requests. The expression "functional code" denotes any code that is productive and that resides in the behaviors.

Reaction steps are the means by which the component adapts itself. It can be for example the replacement of the active behavior, the tuning of some parameters, or the redistribution of arrays. These pieces of code are dynamically inserted in the execution flow when the component adapts itself. Reactions must ensure that they leave the component in such a consistent state that the execution can resume and lead to the same result than if no reaction has been executed.

Because reactions must enforce the consistency of the component, reaction steps cannot be inserted at any time in the execution flow. In order to specify the places at which the component is able to adapt itself safely, we define the notion of *adaptation point*.

5.2   Semantics of Adaptation Points

An adaptation point is an annotation in the code that indicates where the component can be safely modified. The developer indicates that the behavior is able to suspend in a consistent state and to resume from this state at an adaptation point, no matter which behaviors and reaction steps combination leads the component to that state.

The platform enforces the mutual exclusion between the functional code and reactions: it ensures that reactions might only be executed when the functional code is suspended at an adaptation point. Adaptation points are thus the moments at which reaction steps can be inserted in the execution flow.

Reaction steps must ensure that if the state of the component is consistent, it remains consistent after the execution of the whole reaction. The scope of this consistency includes both the variables of the component and the active behavior. This means that the functional code should not be able to determine whether a particular reaction step has been executed at a particular adaptation point. Adaptation points are almost invisible to the functional code.

An adaptation point is said *active* when a reaction is scheduled at that adaptation point. Otherwise, the adaptation point is said *inactive*.

5.3 Introducing Global Adaptation Points

Because several processes may collaborate during a single reaction, they need to be synchronized and coordinated. As for global consistent states in distributed systems, some of the combinations of adaptation points do not represent valid states at which the component is able to adapt itself.

Adaptation points are local to each process; so are the annotated states. For that reason, adaptation points are not sufficient to specify states at which the whole behavior can be modified. This is why the developer has to give explicitly a compatibility relationship between the adaptation points of each process of the behavior in order to allow the platform to find consistent states. The platform enforces the fact that reactions can only be executed when all the processes are suspended on adaptation points that are compatible with each other. Those *global adaptation points* specify global states at which the component is able to adapt itself, global states at which the developer permits the adaptation mechanism to be executed. Our model only specifies the semantic of global adaptation points: the developer should place them to indicate the global states at which he thinks the adaptation can occur safely.

5.4 Building a Host Platform

We consider that modern programming techniques should conform to the "separation of concerns" paradigm. For this reason, we think that adaptability should be a service given by the platform that hosts the component is deployed. Figure 2 shows the overall architecture of a platform hosting a parallel self-adaptable component.

The platform mainly provides two kinds of objects: the *decider* and the *coordinators*. The decider is the object that makes the decisions: it decides when

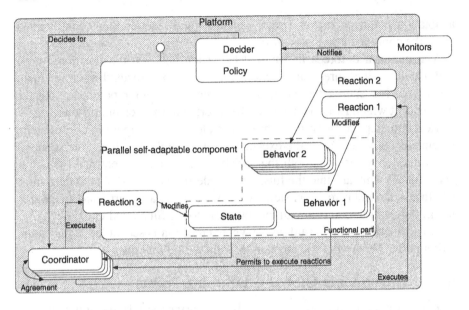

Figure 2.   Overall architecture of a parallel self-adaptable component

(events to watch) and how (reactions to execute) the component should adapt itself according to the adaptation policy. It bases its decisions on the reports given by the *monitors* that are interfaced with the platform. The monitors are daemons that track and report changes in the state of the environment. The *coordinators* execute the directives given by the decider: they serve as intermediaries between the code of the component and the platform. Their role is to synchronize the adaptation mechanism with the functional code and to coordinate the execution of the reactions.

When a monitor detects a change in the state of the resources it watches, it notifies the decider. The decider then interprets the given adaptation policy. During this step, the decider might query the monitors for a detailed report on the state of the resources. If it concludes that there is no need to adapt, it stops the adaptation process until the next notification from a monitor; otherwise, it broadcasts its decision to the coordinators, which decide collectively when the reaction is effectively executed. This is done by selecting a global adaptation point - and activating the corresponding adaptation point of each process. Once a process reaches an active adaptation point, its coordinator executes the reaction chosen by the decider.

6.   Experimental Results

In order to validate the feasibility of our approach, we have built a prototype platform that partially implements the model we have proposed.

6.1   Test Platform

The test platform we developed implements a subset of the model. It concentrates on SPMD applications. The adaptation points we implemented allow to execute user-defined reaction.

Although the application we use in our tests seems simple, it is sufficient to show that our adaptation model works well with parallel applications without significant performance degradation. Our application is a generic vector iteration that distributes its vectors using a block scheme. It uses MPI for its communications. Its adaptation policy is to use as many nodes as the monitor reports: it spawns new processes when nodes are added to the system and terminates the processes that use nodes reclaimed by the system. Because the MPI implementation we use cannot dynamically spawn or terminate processes, the application starts with a fixed number of processes, but uses only the reported number. For this reason, we simulate the monitoring of the number of available processors; this also allows us to have full-control over the adaptation mechanism. We place one adaptation point between iterations.

6.2   Gain of the Adaptation

In this experience, once the application has been started, the number of available nodes is increased from four to six. Figure 3 shows the elapsed time at the end of the iterations. The execution of the reaction occurs between iterations 12 and 13; this appears as a break on the curve. This figure shows that several iterations are needed in order to balance the cost of the execution of the reaction. In the long term, the gain of the adaptable version over the original application is substantial. The amount of iterations needed before the adaptable version becomes effectively better depends on the reaction that has been executed and on the component itself.

6.3   Overhead of the Adaptation Platform

We compared the execution time of the original application and of its adaptable version in a totally static environment. The difference linearly depends on the number of encountered adaptation points. It shows the time needed to initialize the platform and the execution time lost in adaptation points. The execution time overheads are shown in Table 1, depending on the number of encountered adaptation points. This measure is very noisy. The execution time lost in each adaptation point seems constant at about 0.3 seconds.

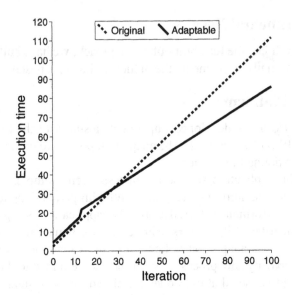

Figure 3.   Execution time of an adaptable application (in seconds)

Table 1.   Execution time for several iteration counts

Adaptation points	Overhead (ms)
50	264.9
100	295.6
150	304.3
200	376.4
250	278.5

6.4   Ease of Use

Our prototype assumes that the application is iterative and that adaptation points only reside between iterations. The reason is that it makes it easier to build the state of spawned processes. However, our model does not impose such a restriction.

One of the hardest questions to answer to when using the platform is the placement of adaptation points and their frequency. For our experiments, we arbitrarily place one adaptation point per iteration, but there was no a priori reason for doing so. Having many adaptation points makes the application more reactive to environmental changes at the cost of an increased overhead.

7. Conclusion

In this chapter, we have shown that the idea of combining a dynamic adaptation framework with parallelism and distribution is a promising way for efficiently programming the Grid. We have built a prototype that extends the ACEEL adaptation engine to take into account the parallelism that can reside in components. This allowed us to experiment our ideas.

Our model complies with the separation of concerns paradigm, since it completely separates the adaptation mechanism from the functional code of the component. Moreover, it provides a basis for the dynamic adaptation of parallel code, whereas many projects have focused on their configuration at startup. This separation allowed us to experiment our idea without integrating a full-featured component platform. However, we expect that building application using a component infrastructure help doing the adaptation. Indeed, containers offered by the component infrastructures are a privileged place into which non-functional services (such as security or adaptation) should reside. Those containers also help doing the adaptation with their introspection ability.

In our ongoing works, we plan to define more formally the properties that the component is required to satisfy in order to be able to adapt itself. This includes the properties of global states where an adaptation can occur. The constraints on behavior replacement should also be investigated. The goal of those studies is to help the developer when establishing constraints on the adaptation. We expect this will allow to detect automatically valid adaptation points or at least to check that the points specified by the developer are correct.

Studying the relationship between fault tolerance systems that use checkpointing and adaptation in the context of Grid computing is an important perspective. Firstly, finding shared properties between checkpoints and adaptation points would be of great help in establishing properties and constraints on adaptation point placement. Secondly, fault tolerant systems suppose that the execution environment is dynamic since any fault results in changes in the environment. Keeping in mind fault-tolerance related works seems a natural approach since they share with adaptation the need to find "special points" at which the execution can resume. However, fault tolerance systems try to repair faults rolling-back the execution, whereas adaptation does its best until the code is able to react.

By the time, we have only studied the overall architecture for the adaptation of parallel codes. Our work is currently focusing on how to choose the global adaptation point at which the reaction steps should be executed. This work is going to lead to the definition of an algorithm to suspend the execution in a well-defined point without rollback.

We have no precise idea of the overhead required from the developer to make a component able to adapt itself using a generic framework. However, several examples exist of parallel codes made dynamically adaptable in an ad-hoc way and we expect that having a generic framework simplifies the task of the developer. In order to have a reasonable opinion about this subject, we plan to study how several parallel and distributed codes can be made self-adaptable. In particular, we think of how parallel discrete event simulators can be modified to adapt itself to the execution environment. We expect from these experiments to get a measurement of the work such a goal needs.

References

[1] D. Chefrour and F. André. Développement d'applications en environnements mobiles à l'aide du modèle de composant adaptatif aceel. In *Langages et Modèles à Objets LMO'03. Actes publiés dans la Revue STI, série L'objet, volume 9*, Vannes, France, February 2003 (in French).

[2] A. Denis, C. Pérez, and T. Priol. Portable Parallel CORBA Objects: An Approach to Combine Parallel and Distributed Programming for Grid Computing. In *Proc. of the 7th Intl. Euro-Par'01 Conference (EuroPar'01)*, LNCS, 2150:835–844, Springer, August 2001.

[3] I. Foster, C. Kesselman, J. Nick, and S. Tuecke. The Physiology of the Grid: An Open Grid Services Architecture for Distributed Systems Integration. In *Global Grid Forum*, June 2002.

[4] E. Gamma, R. Helm, R. Johnson, and J. Vlissides. *Design Patterns: Elements of Reusable Object-oriented Software*. Addison Wesley, 1998.

[5] Globus Toolkit. http://www.globus.org.

[6] E. Jeannot, B. Knutsson, and M. Björkman. Adaptive Online Data Compression. In *11th IEEE International Symposium on High Performance Distributed Computing (HPDC-11 2002)*, pages 379–388. IEEE Computer Society, 2002.

[7] K. Keahey and D. Gannon. PARDIS: A parallel Approach to CORBA. In *HPDC*, pages 31–39, 1997.

[8] X. Martorell, J. Corbalán, N. Navarro, and J. Labarta. The NANOS Resource Management System. In *4th Operating System Design and Implementation (OSDI 2000)*, 2000.

[9] Object Management Group. Corba Components, June 2002. Document formal/02-06-65.

[10] C. Pérez, T. Priol, and A. Ribes. A Parallel CORBA Component Model for Numerical Code Coupling. In *Proc. 3rd International Workshop on Grid Computing*, LNCS, 2536:88–99, Springer, November 2002.

[11] C. Szyperski. *Component Software: Beyond Object Oriented Programming*. Addison Wesley, 1998.

[12] *Wireless Application Protocol 2.0: Technical White Paper.* http://www.wapforum.org, January 2002.

HOCS:
HIGHER-ORDER COMPONENTS FOR GRIDS

Martin Alt, Jan Dünnweber, Jens Müller, and Sergei Gorlatch
Institut für Informatik
University of Münster
Germany
mnalt@math.uni-muenster.de
duennweb@math.uni-muenster.de
jmueller@math.uni-muenster.de
gorlatch@math.uni-muenster.de

Abstract   We present *HOCs – Higher-Order Components –* that provide the Grid application programmer with reusable and composable patterns of parallelism. HOCs can be viewed formally as higher-order functions, i.e. a generic implementation of a HOC on a remote machine can be customized with application-specific code parameters which are supplied by the user and shipped via the network. We take the well-known "Farm of Workers" pattern as our motivating example, present an experimental implementation of the Farm-HOC as a Grid Service using the Globus Toolkit, and report first measurements for a case study of computing fractal images using the Farm-HOC.

Keywords: Grid services, WSRF, Globus toolkit, code mobility

1.   Introduction

This chapter addresses the important and difficult problem of simplifying the programming of Grid applications. Our foremost goal is a high-level programming model which would shield the application programmer from the low-level details of heterogeneous and highly dynamical Grid environments, thus allowing to concentrate on algorithmic and performance issues.

The presented work in progress originates from the following two veins of previous research: 1) Component-based software development [11], with the major goals of reuse and compositionality, and 2) Skeleton-based programming [2], aiming at identifying and abstracting typical patterns of parallel computing. We introduce a new kind of components – *HOC (Higher-Order Components)*, which can be parameterized with application-specific code – and discuss both their use and implementation in the contemporary Grid context of OGSA/WSRF [3] and the Globus Toolkit [6].

The particular contributions and structure of the chapter are as follows. Section 2 motivates and introduces HOCs and explains their use in a Grid environment. In Section 3, we describe an exemplary implementation of the Farm-HOC using the Globus Toolkit. Section 4 presents an application case study of computing fractal images, programmed using the Farm-HOC, and reports first experimental results on our Grid testbed. We compare our results to related work in Section 5.

2.   HOCs: Motivation and Use for Grids

Components, as described by e.g. [11], capture common programming patterns as independent, composable program units and present a high-level API to the application programmer, hiding hardware specific details. Among other benefits (e.g. separation of concerns), an important advantage of using components is code reuse: different applications requiring common functionality can share a common component implementing that functionality.

Our motivation for *higher-order components (HOCs)* is that components for parallel computing often need to be parameterized not only with data but also with application-specific code. An example for such a component is the *Farm* pattern of parallelism, where an application problem is divided into several independent subtasks, which are computed by different *Workers* in parallel. This pattern is very common, e.g. for image processing applications, where an image is divided into several segments and each segment is processed independently by a worker. Instead of re-implementing this pattern for different image processing algorithms, it would be more efficient to provide a generic Farm-HOC, which implements the distribution of the input data across processors, the initialization of workers and finally collecting the results in an application-independent manner. The code for the image processing by the workers can be provided by the application programmer as an application-specific parameter when the Farm-HOC is invoked.

2.1 Grid Programming Using HOCs

The idea of Grid application programming using HOCs is illustrated in Figure 1. The hosts in the Grid provide architecture-tuned libraries of time-intensive HOCs and offer them in the form of Grid Services. An application developer expresses the application in terms of the available HOCs and implements the necessary code parameters (denoted $A, B, \ldots$ in the figure). To make the application-specific parameters available to the hosts, the parameters are then stored to a code base (step ① in the figure) accessible by the hosts.

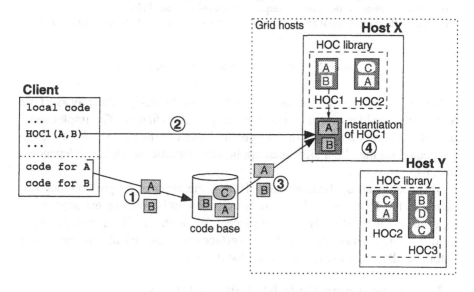

Figure 1. Grid application programming using HOCs

When the application is executed on the client, the used HOCs are called remotely on high-performance Grid hosts (step ② in the figure: the client calls component HOC1 with code parameters A and B); application-specific data and references to code parameters in the code base are passed as arguments. The host checks if the application specific code parameters are available locally and, if they are not, retrieves the code from the code base (③). Once the application specific code is available on the host, it is linked to the HOC implementation in the library and the instantiation of the HOC with the provided code parameters is executed (④).

The implementation of a HOC consists of two parts: a) a set of interfaces specifying the signatures of the HOC's parameters, and b) server-sided interface implementations. Since code parameters for a HOC carry method implementations, the signature of each parameter (method's name, types of the arguments and the result) must be provided by the HOC developer to enable the method for remote invocation.

Figure 2 shows simplified interfaces for the Farm-HOC in Java notation. There are two code parameters: 1) the Master that splits the input data in an application-specific manner for distribution among the workers, and 2) the Worker that processes a unit of data in an application specific manner.

```
interface FarmHOC {
  setMaster(int masterID); setWorker(int workerID);
  double[] compute(double[] input);}
interface Worker { double[] compute(double[] input);}
interface Master { double[][] split(double[] input, int numWorkers);}
```

Figure 2.   Simplified interfaces for the Farm-HOC.

The FarmHOC interface is implemented in a hardware-specific way as a Grid Service that resides in a particular Grid host's HOC-library. The implementation of the Farm-HOC uses the Master and Worker interfaces, which are implemented on the client side in an application specific, hardware-independent manner.

To develop an application using HOCs, the application programmer first identifies the HOCs suitable for the application and expresses the application in terms of the selected HOCs' interfaces. When the application is executed, Grid hosts implementing the HOC interfaces are selected at runtime (either automatically or by the user), using a lookup service.

2.2   Introducing Code Mobility to OGSA

Because HOCs are implemented on Grid hosts while the application-specific customizing code (parameters) reside on clients, HOCs require facilities for code mobility, i.e. shipping code from clients to servers and executing it there. Code mobility mechanisms are currently available in Java/RMI based distributed computing technologies like Jini or JXTA. However, the communication protocols used by RMI are often unable to pass through firewalls and Internet-proxies in a Grid environment. Additionally, using RMI for communication would reduce our language choice to Java. In contrast, Grid Services, as proposed by OGSA/OGSI, are suitable for Internet-wide applications, allowing communication across firewalls and proxies. Grid Services use standard XML-languages for remote procedure calls and interface descriptions, namely SOAP and WSDL. Therefore, it is possible to combine Grid Services developed in different programming languages.

Since OGSA does not define any standard implementation for code mobility, we propose to extend the Grid Services design suggested by OGSA/WSRF by a corresponding mechanism. In comparison to RMI, the SOAP mechanism used by Grid Services is more restrictive: it is not possible to pass an object of any type that cannot be expressed using an XML-schema. Therefore, SOAP parameters may consist of primitives or some kind of data record that may be declared in a class-like manner, but code in form of a method implementation cannot be transferred.

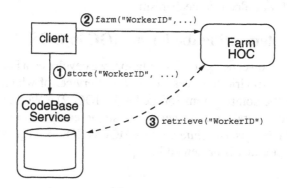

Figure 3. Code mobility mechanism for enabling HOCs

We implement code mobility in OGSA in the following way (shown in Figure 3 for the Farm-HOC). The application programmer stores the implementation of code parameters for HOCs to a code base, and the HOC implementation on the server retrieves the implementation later from there. The code base itself is integrated into the system as a Grid Service providing a `store` method for the clients and a `retrieve` method for servers. When a client stores code to the code base, it uses a unique name (analogous to the Grid Service Handles used in Globus) to identify the stored code. This name ("WorkerID" in the figure) is then passed to the HOC as a parameter upon invocation. The HOC implementation uses this name to retrieve the code from the code base service. Finally, the code is linked to the server sided HOC implementation and the execution of the HOC can be initiated.

The proposed code mobility mechanism assumes that the code for a HOC parameter is executable on the HOC host. This can be achieved either by using portable code for the parameters (e.g. Java Bytecode or an interpreted language such as Python), or by sending the parameters as source code (e.g. C or C++) and compiling them to machine code on the HOC-host. This requires an advanced infrastructure, capable of handling compilation errors at runtime, etc. We have recently developed a prototype that allows for remote compilation of code parameters via a portal; a detailed description of this feature is in preparation. Another possibility is to compile the parameter for the HOC-host's processor type on the client, using a cross-compiler, and to store the machine code to the code base.

For reasons of simplicity we have chosen to use Java in our experimental implementation. For server-sided HOCs, we implemented a code loading mechanism based on a remote class loader, which replaces the Java default class loader when needed and connects to the code base service instead of searching class files on the local hard disc when new classes are loaded. Once the bytecode for a particular class is retrieved from the code base, the class is instantiated using the Java reflection mechanism.

3.   Implementing the Farm-HOC in OGSA

As a proof-of-concept implementation, we developed a Farm-HOC as a combination of two Grid Services: 1) a Farm Service, of which a single instance distributes the computations for the Farm-HOC. 2) a Worker Service, which is instantiated several times to perform a single computation. The detailed discussion of the *service architecture* for HOCs (HOC-SA) goes beyond the scope of this chapter and is presented in [5].

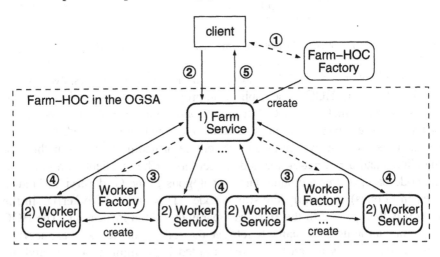

Figure 4.   Farm-HOC processing in OGSA

Figure 4 shows the computation and communication steps in the OGSA-based Farm-HOC processing. The Client firstly obtains a Farm Service instance from the so-called HOC Factory (①) which exploits the factory design pattern in the typical OGSA-like manner. To start calculations, the client communicates with the Farm Service (②) which is a kind of facade for the whole Farm-HOC. The Farm Service obtains the Worker Services from the corresponding Factories (③), creates and distributes the sub-tasks of calculation (④) and reassembles the result which finally is returned to the client (⑤). This way, the Farm Service hides all the distributed interactions in the Grid and provides a lean interface for the complete Farm-HOC.

Parallelism in the Farm-HOC is achieved by starting a new thread (taken from a thread pool) for each worker, which sends back a notification to the Farm Service upon completion of the calculation.

An advantage of HOCs is that all the cumbersome implementation work in Globus needs to be done only once for each HOC. despite the simplicity of the farm-hoc, its implementation using the currently available Globus Toolkit has required several laborious steps, which are difficult for an application programmer: 1) defining remote interfaces for the Farm Service and the Farm-Worker Service using a Globus specific configuration file (GWSDL file), 2) implementing the services using the Globus Toolkit APIs, 3) writing a deployment descriptor in another Globus specific configuration file (WSDD file) 4) creating an archive (GAR), containing the services and configuration files in a Globus-specific hierarchy, 5) setting up a runtime environment at all remote hosts and deploying the services there. It is an advantage of our approach that these steps are not done repeatedly by application programmers, but rather accomplished once for a HOC.

```
farmHOC = farmHOCFactory.createHOC();
farmHOC.setMaster("masterRef"); farmHOC.setWorker("workerRef");
farmHOC.configureGrid( "masterHost",
                       "workerHost1",... , "workerHostN" );
farmHOC.compute(input);
```

Figure 5. Farm Customization and Invocation Using the Farm-HOC

Figure 5 shows how the application programmer uses the Farm-HOC. The application starts by requesting a new instance of the Farm-HOC from the FarmHOC factory (line 1). Line 2 passes references to the application-specific code parameters (as specified by the interfaces in Figure 2) in the code base to the HOC-instance. In lines 3 and 4, the hosts for both master and workers are selected. The HOC is then invoked to process input data on the Grid (line 5).

4. Case Study: Calculating Julia Sets

As an application of our Farm-HOC, we calculate so-called Julia Sets, a special kind of fractal image. Julia Sets are computed by applying a given function to each point of the complex plane (within a certain range). This function is then iterated to produce a sequence of complex numbers that may diverge or converge; the sequence's degree of growth determines the color assigned to a particular point in the plane. The computation is a dynamic process that requires different amounts of time for different points.

4.1   Using Farm-HOC

The calculation process can be applied to each point independently, which allows a straightforward parallelization by dividing the plane into p rectangles and distributing the computations among p processors. To implement this schema using our Farm-HOC, the application developer needs to provide a master and a worker implementation (the latter is shown in Figure 6).

```
public class FractalWorker implements Worker  {
  public Object compute(Object tile) {
    for (int y = 0;  y < tileHeight;  ++y) {
      for (int x = 0;  x < tileWidth;  ++x) {
      ...  // compute julia value for (x, y) }}...}
```

Figure 6.   Example Worker Parameter for the Farm-HOC

The code shown in Figure 6, which is uploaded and stored in the code base, is used to instantiate the Farm-HOC in line 2 of Figure 5.

Our Farm-HOC implementation allows for a simple self-scheduling strategy. A Worker is implemented using multiple threads taken from a thread pool, so it can process parts of the input in parallel itself. Now, if the tasks are unequally balanced on a compute node, a worker thread that has been suspended early will immediately free resources for the thread which is processing a more time-intensive task.

4.2   Experimental Results

Table 1 shows the results of preliminary tests conducted with the Farm-HOC. Our experimental Grid testbed consists of one host in Münster running the master implementation, and up to three remote multiprocessor hosts in Berlin (at a distance of 500 km), each running multiple parallel workers. The underlying TCP/IP network has the bandwidth of 1 Mb/s and the latency of 25 ms.

Table 1.   Performance measured for Farm-HOC.

1 remote server (4 processors)	2 remote servers 4 + 8 (processors)	3 remote servers (4 + 8 + 12 processors)
198,212 sec	128,165 sec	48,377 sec

The server in Münster was a Linux PC (Pentium 4 running at 2.6GHz) and the remote servers were SunFire multiprocessors (running Solaris with 4, 8, and 12 processors correspondingly). For Julia Sets, all tasks have different time costs and the compute hosts have different computing power, so the scaling behavior is not regular. Nevertheless, the results show that the application does scale. The variations in multiple measurements were low. The sequential time of a local evaluation on the PC was more than five times higher than using a remote server with 4 processors.

Another result of experiments was that transferring the result via SOAP takes much time (about 60 sec), due to the complexity of the SOAP encoding. We plan to exploit GridFTP using Java CoGs [13] to reduce this time.

5. Related Work

Our work is one of the current efforts on hiding the complexity of Grid-aware application development (a good overview can be found in [7]). Related work includes the Ibis programming system [12], the skeleton library Lithium [4], and the ProActive [1] system, all three based on Java/RMI, and the PaCO++ library based on the CORBA communication mechanism [9]. Another approach in providing a high-level programming interface to the Grid application developer, that does not have its origins in skeletal programming is GridRPC [10]. The GridRPC API is composed of a collection of lightweight C-functions, to perform simple remote procedure calls in Grid systems like Ninf or NetSolve. They communicate using non-XML proprietary protocols that are partially more efficient than SOAP though not OGSA-compliant. Our object-oriented HOCs are designed to be used in the OGSA, so they use Java together with Globus to ensure interoperability. Higher-order components allow to free the application programmers from much of the complexity when using the current version of the Globus Toolkit implementation described e.g., in [8].

6. Conclusions

The higher-order components (HOCs) proposed in this chapter provide the application programmer with a high-level programming interface for OGSA-based Grid programming. They ensure the separation of concerns between the system programmers who develop and implement HOCs and the application programmers who use HOCs by instantiating them with code parameters. The implementation of the Farm-HOC has demonstrated that such components are reusable in different applications and offer promising performance. We are in the process of introducing and implementing new HOCs. They will include components that encapsulate task-parallelism and also communication between nodes, rather than the simple completion notifications as in the Farm-HOC.

References

[1] L. Baduel, F. Baude, and D. Caromel. Efficient, Flexible, and Typed Group Communications in Java. In *Proceedings of ACM JavaGrande ISCOPE 2002 Conference*, pages 28–36, ACM Press, November 2002.

[2] M. I. Cole. *Algorithmic Skeletons: A Structured Approach to the Management of Parallel Computation*. Pitman, 1989.

[3] K. Czajkowski, D. Ferguson, I. Foster, J. Frey, S. Graham, I. Sedukhin, D. Snelling, S. Tuecke, and W. Vambenepe. The WS-Resource Framework, May 2004. http://www.globus.org/wsrf/.

[4] M. Danelutto and P. Teti. Lithium: A Structured Parallel Programming Enviroment in Java. In *Proceedings of Computational Science - ICCS*, LNCS, 2330:844–853, Springer, April 2002.

[5] J. Dünnweber and S. Gorlatch. HOC-SA: A Grid Service Architecure for Higher-order Components. In *Proceedings of the 2004 IEEE International Conference on Services Computing (SCC 2004), Shanghai, China*. IEEE Computer Society Press, September 2004 (to appear).

[6] Globus Alliance. Globus Toolkit http://www.globus.org/toolkit/.

[7] D. Laforenza. Grid Programming: Some Indications Where We Are Headed. *Parallel Computing*, 28(12):1733–1752, December 2002.

[8] G. Mair and A. Villazón. Implementing a Distributed Master/Slave Grid Service with Globus Toolkit 3 (GT3). http://dps.uibk.ac.at/~gregor/mandel.pdf.

[9] C. Pérez, T. Priol, and A. Ribes. PaCO++: A Parallel Object Model for High Performance Distributed Systems. In *Proceedings of the Conference on System Sciences (HICSS-37)*. IEEE Computer Society Press, January 2004.

[10] K. Seymour, H. Nakada, S. Matsuoka, J. Dongarra, C. Lee, and H. Casanova. Overview of GridRPC: A Remote Procedure Call API for Grid Computing. In M. Parashar, editor, *Grid Computing - GRID 2002, Third International Workshop*, LNCS, 2536:274–278, Springer, November 2002.

[11] C. Szyperski. *Component Software: Beyond Object-oriented Programming*. Addison Wesley, 1998.

[12] R. V. van Nieuwpoort, J. Maassen, R. Hofman, T. Kielmann, and H. E. Bal. Ibis: an Efficient Java-based Grid Programming Environment. In *Proceedings of ACM JavaGrande ISCOPE 2002 Conference*, pp. 18–27. ACM Press, November 2002.

[13] G. von Laszewski, I. Foster, J. Gawor, and P. Lane. A Java Commodity Grid Kit. *Concurrency and Computation: Practice and Experience*, 13(8-9):643–662, 2001.

THE COMPONENT ARCHITECTURE OF OPEN MPI: ENABLING THIRD-PARTY COLLECTIVE ALGORITHMS*

Jeffrey M. Squyres and Andrew Lumsdaine
Open Systems Laboratory
Indiana University
Bloomington, Indiana, USA
jsquyres@open-mpi.org
lums@open-mpi.org

Abstract   As large-scale clusters become more distributed and heterogeneous, significant research interest has emerged in optimizing MPI collective operations because of the performance gains that can be realized. However, researchers wishing to develop new algorithms for MPI collective operations are typically faced with significant design, implementation, and logistical challenges. To address a number of needs in the MPI research community, Open MPI has been developed, a new MPI-2 implementation centered around a lightweight component architecture that provides a set of component frameworks for realizing collective algorithms, point-to-point communication, and other aspects of MPI implementations. In this chapter, we focus on the collective algorithm component framework. The "coll" framework provides tools for researchers to easily design, implement, and experiment with new collective algorithms in the context of a production-quality MPI. Performance results with basic collective operations demonstrate that the component architecture of Open MPI does not introduce any performance penalty.

Keywords: MPI implementation, parallel computing, component architecture, collective algorithms, high performance

---

*This work was supported by a grant from the Lilly Endowment and by National Science Foundation grant 0116050.

1.   Introduction

Although the performance of the MPI collective operations [6, 17] can be a large factor in the overall run-time of a parallel application, their optimization has not necessarily been a focus in some MPI implementations until recently [22]. MPI collectives are only a small portion of a production-quality, compliant implementation of MPI; implementors tend to give a higher priority to reliable basic functionality of all parts of MPI before spending time tuning and optimizing the performance of smaller sub-systems.

As a direct result, the MPI community has undertaken active research and development of optimized collective algorithms. Although design and theoretical verification is the fundamental basis of a new collective algorithm, it must also be implemented and used in both benchmark and real-world applications (potentially in a variety of different run-time / networking environments) before its performance can be fully understood. The full cycle of design, development, and experimental testing allows the refinement of algorithms that is not possible when any of the individual steps are skipped.

1.1   Solution Space

Much research has been conducted in the area of optimized collective operations resulting in a wide variety of different algorithms and technologies. The solution space is vast; determining which collective algorithms to use in a given application may depend on multiple factors, including the communication patterns of the application, the underlying network topology, and the amount of data being transferred. Hence, one set of collective algorithms is typically not sufficient for all possible application / run-time environment combinations. This is evident in the range of literature available on different algorithms for implementing the MPI collective function semantics.

It is therefore useful to allow applications to select at run-time which algorithms are used from a pool of available choices. Because each communicator may represent a different underlying network topology, algorithm selection should be performed on a per-communicator basis. This implies that the MPI implementation both includes multiple algorithms for the MPI collectives and provides a selection mechanism for choosing which routines to use at run-time.

1.2   Implementation Difficulties

There are significant barriers to entry for third-party researchers when implementing new collective algorithms. For example, many practical issues arise when testing new algorithms with a wide variety of MPI applications in a large number of run-time environments. To both ease testing efforts and to make the testing environment as uniform as possible, MPI test applications should be able to utilize the new algorithms with no source code changes. This will even allow real world MPI applications to be used for testing purposes; the output and performance from previous runs (using known correct collective algorithms) can be compared against the output when using the collective algorithms under test.

This means that functions implementing new algorithms must use the standard MPI API function names (e.g., MPI_Barrier). Techniques exist for this kind of implementation, but they may involve significant learning curves for the researcher with respect to the underlying MPI implementation: how it builds, where the collective algorithms are located in the source tree, internal restrictions and implementation models for the collective functions, etc.

1.3   A New Approach

To address a number of needs in the MPI research community, Open MPI [5] has been developed; a new MPI-2 implementation based upon the collected research and prior implementations of FT-MPI [3–4] from the University of Tennessee, LA-MPI [1, 7] from Los Alamos National Laboratory, and LAM/MPI [2, 19] from Indiana University. Open MPI is centered around a lightweight component architecture that provides a set of component frameworks for realizing collective algorithms, point-to-point communication, and other aspects of MPI implementations.

In this chapter, we focus on the collective algorithm component framework. The "coll" framework provides tools for researchers to easily design, implement, and experiment with new collective algorithms in the context of a production-quality MPI. Collective routines are implemented in standalone components that are recognized by the MPI implementation at run-time. The learning curve required to create new components is deliberately small to allow researchers to focus on their algorithms, not the details of the MPI implementation. The framework also offers other benefits: source and binary distribution of components, seamless integration of all algorithms at compile and/or run-time, and fine-grained run-time selection (on a per-communicator basis).

This chapter is therefore not about specific collective algorithms, but rather about providing a comprehensive framework for researchers to easily design, implement, and experiment with new collective algorithms. Components containing new algorithms can be distributed to users for additional testing, verification, and finally, production usage.

Both MPICH and MPICH2 [8–9] use sets of function pointers (to varying degrees) on communicators to effect some degree of modularity, but have no automatic selection or assignment mechanisms, therefore requiring abstraction violations (the user application has to assign function pointers inside an opaque MPI communicator) or manual modification of MPICH itself.

LAM/MPI v7 debuted the first fully-integrated component-based framework that allowed source and binary distribution of several types of components (including collective algorithms) while requiring no abstraction violations or source code changes to the MPI implementation in a production-quality, open-source MPI implementation. Open MPI evolves these abstractions by refining the concepts introduced in LAM/MPI v7, essentially creating a second generation set of component frameworks for MPI implementations called the MPI Component Architecture (MCA) [5, 23]. This chapter presents Open MPI's MCA collective component framework design.

The rest of this chapter is organized as follows. Section 2 discusses the current state of the art with regards to implementing third-party collective algorithms within an MPI framework. Section 3 describes Open MPI's component model for collective algorithms, and explores different possibilities for third-party implementations. Section 4 provides overviews of two collective modules that are included in the Open MPI software distribution. Finally, Sections 5, and 6 discuss run-time performance, final conclusions, and future work directions.

2.   Adding Collective Algorithms to an MPI Implementation

Third-parties implementing new collective functions can encounter both technical and logistical difficulties, even in MPI implementations that encapsulate collective function pointers in centralized locations. Not only is it desirable for MPI applications to invoke new collective routines through the standard MPI API, there must be a relatively straightforward mechanism for making the new routines available to other users (download, compile, install, compile / link against user applications, etc.).

2.1   Common Interface Approaches

Common approaches to developing new collective routines include: using the MPI profiling layer, editing an existing MPI implementation, creating a new MPI implementation, and using alternate function names. Each of these scenarios have benefits and drawbacks, but all require the collective algorithm author to implement at least some level of infrastructure to be able to invoke their functions.

Use the MPI Profiling Layer.   The MPI profiling layer was designed for exactly this purpose: allowing third-party libraries to insert arbitrary functionality in an MPI implementation. This can be done without access to the source code for either the MPI implementation or the MPI application.

This approach has the obvious advantage that any MPI application will automatically use the new collective routines without modifications. Although the

MPI application will need to be relinked against the new library, no source code changes should be necessary. A non-obvious disadvantage is that since the profiling layer uses linker semantics to overload functions, only one version of an overloaded function is possible. For example, MPI_BARRIER cannot be overloaded with both a new collective routine *and* a run-time debugging/profiling interface.

Edit an Existing MPI Implementation.   This method involves editing an MPI implementation to either include new collective routines in addition to the implementation's existing routines [21–22], or outright replacing the implementation's collective routines with new versions [10]. This can only be used with MPI implementations where the source code is available and the license allows such modifications.

Similar to the profiling approach, this method allows unmodified MPI applications to utilize new functionality. This is perhaps the easiest method for MPI applications because the API is the same and the new routines are in the MPI implementation itself.

However, the learning curve to add or replace functionality in the MPI implementation may be quite large. Additionally, editing the underlying MPI effectively creates a "fork" in the implementation's development path. This may make the code difficult to maintain and upgrade.

Create a New MPI Implementation.   Entirely new MPI implementations have been created simply to design, test, and implement new MPI collective algorithms [11–12]. The advantage to this approach is complete control over the entire MPI implementation. This may be desirable for situations where the collective routines are radically different than current MPI implementations allow. For example, PAC-X MPI was created to enable communications in metacomputing environments, requiring alternate collective algorithms for efficiency.

The overhead with this approach is enormous. Writing enough of an MPI implementation such that a simple MPI program that only invokes MPI_INIT, MPI_COMM_RANK, MPI_COMM_SIZE, and MPI_FINALIZE is a monumental task. The time necessary to create an entire MPI framework before actually being able to work on collective algorithms can be prohibitively large.

Use Alternate Function Names.   Perhaps the simplest approach from the algorithm implementor's perspective is to use function names other than the ones mandated by the MPI standard. For example, provide an alternate barrier implementation in the function New_Barrier instead of MPI_Barrier.

Difficulties arise in testing because MPI applications need to be modified to call the alternate functions. This can be as simple as preprocessor macros in a standardized header file, or may entail manually modifying all invocation points in the application. Requiring source code modification necessarily means that precompiled, binary-only MPI applications will not be able to utilize the new functionality.

2.2   A Component-Based Approach

We propose an open, component-based framework for the implementation of collective algorithms that will solve many of the technical and logistical issues faced by third-party collective algorithm researchers. In this framework, a *collective component* is comprised of a set of top-level *collective routines*. A collective routine implements one MPI collective function (such as MPI_BARRIER, MPI_BCAST, etc.). The framework also includes built-in mechanisms for configuration, compilation, installation, and source and binary distributions of components.

The collective component framework was designed and implemented with the following goals:

- Do not require modifying Open MPI source code to import new collective algorithms.

- Allow new collectives to be imported into the MPI implementation at compile- and run-time.

- Provide easy-to-understand interface and implementation models for collective routines that do not require detailed internal knowledge of the MPI implementation.

- Provide minimal overhead before invoking collective routines to maximize run-time performance.

- Allow (but not require) collective routines to be layered upon MPI point-to-point routines.

- Allow collective routines to exploit back-end hardware and network topologies.

- Allow collective components to be layered upon other collective components.

- Facilitate both source and binary distribution of collective components.

- Enable MPI applications to utilize the new collective components without recompiling / relinking.

- Allow multiple collective components to exist within a single MPI process.

- Provide a fine-grained, run-time, user-controlled component selection mechanism.

There are no current plans to allow experimentation with collective algorithms that are not specified by MPI.

3.   Collective Components

Open MPI is based upon a lightweight component architecture, including a component framework for MPI collective algorithms named "coll." The coll component interface was designed to satisfy the goals listed in Section 2.2. coll components can be loaded and selected at compile-time or run-time. For example, multiple coll components are included in the standard Open MPI distribution, but third-party components can also be added at any time.

3.1   Design Overview

The Open MPI component framework manages all coll components that are available at run-time. This management code is typically referred to as the Open MPI coll framework in the discussion below.

Simply put, a coll components is essentially a list of top-level function pointers that the Open MPI infrastructure selectively invokes upon demand. When paired with a communicator, a component becomes a *module* [20]. Top-level MPI collective functions have been reduced to thin wrappers that perform error checking before invoking back-end coll module implementation functions. One coll module is assigned to each communicator; this module is used to implement all MPI collectives that are invoked on that communicator. For example, MPI_BCAST simply checks the passed parameters for errors and then invokes the back-end broadcast function on its assigned coll module.

3.2   Implementation Models

Components are free to implement the standardized MPI semantics in any way that they choose. Most, however, use one or more of the following models: layered over point-to-point, alternate communication channels, or layered over another coll components.

Layered over Point-to-Point.   A simple implementation model is to utilize MPI point-to-point functions to send data between processes. For example, using MPI_SEND and MPI_RECV to exchange data is both natural and easy to understand, freeing the coll component author to concentrate on the components' algorithms and remain independent of how the underlying communication occurs. This model has been used extensively by MPI implementations [8, 19] and third-party collective algorithm researchers [13–14].

Node 0   Node 1

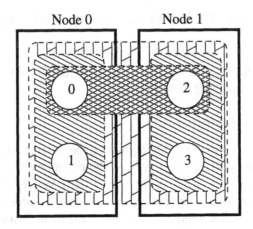

Figure 1. Four processes in MPI_COMM_WORLD are distributed across two nodes. Three sub-communicators (vertical and horizontal) each contain the two processes local to their respective nodes. One "bridge" communicator (horizontal) contains a representative process from each node.

Alternate Communication Channels. Recently, researchers have been exploring the possibility of avoiding MPI point-to-point functionality and instead using alternate communication channels for collective communications. Some network interfaces contain native primitives for collective operations and/or streamlined one-sided operations which can lead to significant performance gains as compared to using traditional point-to-point methods. Examples of alternate communication channels that at least partially support collective operations include (but are not limited to): shared memory [16], UDP multicast [11], Myrinet [24], and Infiniband [15].

Hierarchical coll Components. The coll framework was carefully designed such that coll components can be re-used at run-time in two ways. First, the coll component "basic," as its name implies, is a basic implementation of all of the MPI collectives. It can be used with any communicator and topology. The purpose of this component is to provide a baseline implementation for all MPI collective operations, allowing other components to use its routines as necessary. For example, a component that only provides an optimized scatter algorithm implementation can complete itself by using the methods from the basic component (or other components) for all other collective routines. This allows the optimized scatter component to be used in any MPI program even though it only implements a small number of new/optimized routines.

A second, more complex model involves using a hierarchy of coll modules to implement a single, top-level MPI collective. This is useful when a collective is invoked on a communicator that spans multiple types of networks. For example,

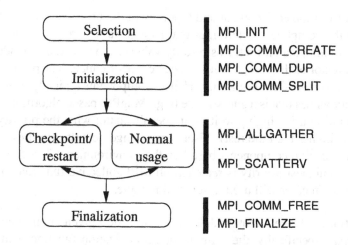

Figure 2. Five phases in the life of a coll component. The component is selected and initialized when a communicator is created. It is used and/or checkpointed during the run, and finalized when the communicator is destroyed.

Figure 1 shows two SMPs, each running two MPI processes. A single MPI communicator contains all four processes. The top-level communicator's coll module creates three sub-communicators: one for each SMP (containing the two processes on each node), and a third "bridge" communicator connecting one representative process from each node.

Note that each sub-communicator will have its own coll module. This hierarchical arrangement of communicators allows each network to utilize its own optimized coll component, resulting in an efficient movement of data across each medium. This model will be explained in more detail in Section 4, where the smp coll component is discussed as an example implementation.

3.3   Component / Module Life Cycle

There are five phases in a coll component's life cycle: selection, initialization, checkpoint / restart, normal operation, and finalization. Figure 2 shows these phases and the corresponding MPI functions that trigger them. Note that a component may be involved in multiple life cycles simultaneously (i.e., several modules of the same component may exist in a single process); coll components have a one-to-many relationship with communicators.

Selection.   As each communicator is created (including MPI_COMM_SELF and MPI_COMM_WORLD), a coll component is *selected* for it from all available components. Specifically, the Open MPI coll framework queries each available coll component to determine if it is available to be used with the newly-created communicator. The queried component analyzes factors such as the current run-time environment and topology of the processes in the communicator. If the component determines that its algorithms are a good match for the target communicator, it returns priority value (from 0 to 100) intended as a relative indicator of the component's expected performance. The priority value is relative and changable at run-time. Hence, components typically provide default priority values that is a guesstimate (e.g., MagPIe-based algorithms across WANs should return a high priority – it doesn't matter what the priority is, as long as it is higher than the rest). Users can change default priorities to force selection of specific components based on their environment. The component returning the highest priority is *selected*; all MPI collective functions invoked on that communicator will use the selected module.

Initialization.   Once a coll module is selected for a given communicator, it is *initialized*. Specifically, the component's initialization function is invoked, passing the target communicator as an argument. The initialization function performs any one-time setup required by the module, and returns a module that contains any local state required to perform collectives on the target communicator. By definition, a communicator's member processes and ordering are static, allowing a module's initialization routine to pre-compute any data structures that will later be used during collective routines. This design emphasizes the potential run-time optimizations that can be obtained by shifting as much overhead calculations and coordination to the one-time initialization function as possible. This can reduce the amount of computational overhead in the run-time of collective routines.

The module is associated with the target communicator by caching its local state (such as the pre-computation results) on the communicator itself. All subsequent phases in the module's life cycle are invoked relative to a communicator for which it was selected; the communicator is passed as an argument to all invocation functions. This allows the module to retrieve its communicator-specific pre-computation data when a collective function is invoked.

Once a component has been initialized, it returns the module – including a list of function pointers for its algorithms – which is then assigned to the communicator. These functions are later invoked by the coll framework during the "normal usage" phase in the module's life cycle whenever a top-level MPI collective function is invoked. The module is then ready to be checkpointed or used for collective operations.

Checkpoint / Restart.   Open MPI includes the capability for parallel MPI applications to be transparently checkpointed and restarted. In order for a parallel MPI application to be checkpointed, all its modules must include checkpoint/restart functionality. Much of this work is usually the responsibility of the point-to-point modules: they must ensure that "in flight" messages will not be lost upon restart. This is typically effected either by draining the network or utilizing acknowledgment / retransmission schemes.

coll modules that are layered on top of MPI point-to-point functionality therefore require no additional work to support checkpoint / restart; all the necessary work is already performed by the point-to-point modules. coll modules that use their own communication channels, however, will typically need to include additional code to support checkpoint / restart functionality. Such modules can provide hook functions that the Open MPI framework will invoke during checkpoints and restarts to perform any required cleanup and re-initialization, respectively.

It is not an error if a module does not include the functionality required for checkpointing and restarting itself; support for checkpoint/restart in a coll module is optional. Currently, the determiniation of whether a process can checkpoint occurs during MPI_INIT: a process is checkpointable only if all the components that may be used in the process support checkpointing (regardless of whether they are selected).

Normal Usage.   After a coll module has been initialized with a communicator, that module's collective routines will be invoked whenever an MPI collective function is invoked on the communicator. Note that since the type of communicator is known at selection and initialization time (i.e., intra- or intercommunicator), it is the module's responsibility to set itself up so that intra- or intercommunicator algorithms are invoked as appropriate.

For example, when the MPI_Bcast C function is invoked on MPI_COMM_WORLD, it checks all of the parameters that are passed into it. It then invokes the the module's broadcast function pointer. The module's broadcast function pointer can either be specifically for intracommunicators or dispatch to an intracommunicator algorithm when it detects the type of MPI_COMM_WORLD. This model allows for a natural separation of algorithms and code since the algorithms used for intracommunicators are, by definition, different than the algorithms used for intercommunicators.

Finalization.   The final phase in a coll module's life cycle on a communicator occurs when the communicator is destroyed. The module's finalization method is responsible for cleaning up all resources associated with the communicator that is being destroyed.

3.4   Component and Module Interfaces

The coll component interface is relatively small; it contains data required for all Open MPI MCA modules such as references to the framework that the component belongs to, the name and version number of the component,

```
coll.component.interface {
  // Metadata identifying what version of the MCA this component
  // adheres to, what framework and version this component belongs to,
  // and this component's name and version.

  version mca_version_number;
  string mca_framework_name;
  version mca_framework_version_number;
  string component_name;
  version component_version_number;

  // Actions defined for all MCA components

  int component_open_function(void);
  int component_close_function(void);

  // Actions defined on coll components.

  int component_init_query(bool &allow_user_threads,
                           bool &have_hidden_threads);
  coll_module component_comm_query(MPI_Comm comm, int &priority);
}
```

Figure 3.   Pseudocode for the coll component interface.

and once-per-process initialization ("open") and finalization ("close") actions. Finally, two actions are defined specifically for coll components:

- One-time initialization. This method is invoked during MPI_INIT to ask certain threading characteristics about the component, and is mainly used to determine the final threading level that will be used during the process (MPI_THREAD_SINGLE through MPI_THREAD_MULTIPLE).

- Per-communicator query. The coll framework invokes this method on each component, effectively asking the component if it wants to be considered for selection. If it does, the component will return a module.

Pseudocode for the component interface is shown in Figure 3.

The module interface is divided into several categories of actions (shown in Figure 4):

- Initialization and finalization. If a module is selected, its initialization method is invoked, allowing the module to complete any setup or precompute results that are utilized during the module's "normal usage" life cycle phase. All modules have their finalize method invoked when they are no longer used (which may be immediately if a module is not selected).

- Checkpoint / restart functionality. As described in [18], the checkpoint/restart functionality in LAM/MPI (and carried forward to Open

```
coll_module_interface {
  // Initialization and finalization of a module

  int init(MPI_Comm comm)
  int finalize(MPI_Comm comm)

  // Checkpoint/restart functionality

  int cr_interrupt(void)
  int cr_checkpoint(MPI_Comm comm)
  int cr_continue(MPI_Comm comm)
  int cr_restart(MPI_Comm comm)

  // Collective algorithm methods

  int allgather(buffer sbuf, int scount, MPI_Datatype sdtype,
                buffer rbuf, int rcount, MPI_Datatype rdtype,
                MPI_Comm comm)
  int allgatherv(buffer sbuf, int scount, MPI_Datatype sdtype,
                buffer rbuf, int rcounts[], int disps[], MPI_Datatype rdtype,
                MPI_Comm comm);
  // ...and the rest of the MPI collective operations
}
```

Figure 4. Pseudocode for the coll module interface. Module-specific state is cached on the communicator and is therefore passed in to every module method.

MPI) consists of three distinct phases: checkpoint, continue, and restart. Methods are included to support each of these actions; their functionality is described further in [18].

- MPI collective functions. Modules contain a method for each MPI collective function (e.g., MPI_BCAST, MPI_BARRIER, etc.). Their function signatures are quite similar to their MPI counterparts, but some of the functions and arguments have been streamlined by the coll framework. For example, some components can treat a zero-byte broadcast as a no-op, and the coll framework will not invoke the module in such situations.

4. Example Components

basic and smp are two of the coll components included in Open MPI. These components serve both as reference algorithms as well as examples of two different implementation models.

4.1 The basic Component

The basic component contains a full set of intra- and intercommunicator collectives. The intracommunicator algorithms are quite mature; they have been in LAM/MPI production code for years. The intercommunicator algorithms are new, but are essentially variations of their intracommunicator counterparts.

Prior generations of LAM/MPI—including the collective algorithms that the basic component is founded on—were based on a monolithic architecture. This made it a natural choice for not only influencing the design of the coll component interface, but also as a first coll component implementation. The successful port of the legacy LAM/MPI collective algorithms to the new framework (originally in LAM/MPI 7.x, and later to Open MPI) served as a validation of the overall coll design.

Although relatively naive, the basic routines can be used on any communicator (regardless of underlying topology), switching between $O(n)$ and $O(log(n))$ algorithms depending on the number of processes in the communicator. All of the basic algorithms essentially use MPI point-to-point functions for moving data between MPI processes. For example, in MPI_BCAST's logarithmic implementation, a traditional binomial tree is used: parent processes send data with MPI_SEND while child processes block in MPI_RECV.

4.2   The smp **Component**

The smp component was also instrumental in shaping the design of the coll framework. Based on the algorithms from the MagPIe project [13–14], the smp algorithms attempt to maximize bandwidth conservation across multiple levels of network latency. MagPIe focused on uniprocessors communicating across a WAN; the smp component is oriented to SMPs communicating on a LAN. The end effect is the same: two levels of network latency that can be exploited at run-time. Segmenting the communicator into groups of local process peers and electing representatives from each group to communicate with other groups provides a natural segregation of local and global communications.

Similar to the basic component, the smp component uses point-to-point communication to pass messages. Standard MPI functions are used to create sub-communicators and translate rank identifications between them. A direct implication of this model is that the coll framework must be able to handle recursive communicator creation and destruction. During the construction of a communicator, the initialization of a coll module may cause the creation of another communicator. This may, in turn, trigger the creation of yet another communicator (and so on).

For example, in the MagPIe broadcast algorithm, the root broadcasts the data to the set of representatives from the other process groups. Each representative (including the root) then broadcasts to the members of its local group (see Figure 5). During the initialization phase of the smp module, the three sub-communicators shown in Figure 5 are created: two containing local-only processes, and one "bridge" communicator between processes 0 and 3. This

Node 0   Node 1

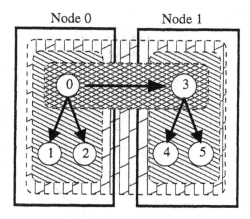

Figure 5. MagPIe algorithm for broadcast from process 0. Process 0 sends to its peer on the remote node (process 3). Each then do a local broadcast to the remaining processes on their nodes (processes 1 and 2, and processes 4 and 5, respectively).

```
int ssi_coll_smp_bcast(buffer, ..., MPI_Comm comm) {
  if (i_am_a_representative) {
    MPI_Bcast(buffer, ..., rep_root, rep_comm);
  }
  MPI_Bcast(buffer, ..., local_root, local_comm);
  return MPI_SUCCESS;
}
```

Figure 6. Pseudocode broadcast implementation using sub-communicators (error handling ignored for this example).

allows the reducing the MagPIe broadcast algorithm implementation to the pseudocode shown in Figure 6.

Note that there are two calls to MPI_BCAST. These broadcasts use whichever module was selected when the sub-communicators were created. Depending on the number of processes and topology involved, the broadcasts may be optimized according to however the selected coll component is implemented.

5. Performance

It is critical that the coll framework does not contribute additional overhead to collective algorithm performance. Measuring this is straightforward: compare the performance of Open MPI's collective functions against the prior generation of LAM/MPI (specifically, v6.5.9) that both provided the algorithms used in the basic component and was based on an integrated, monolithic model.

The collective algorithm implementations used in LAM/MPI 6.5, although somewhat naive, had well-understood behavior characteristics. Its main op-

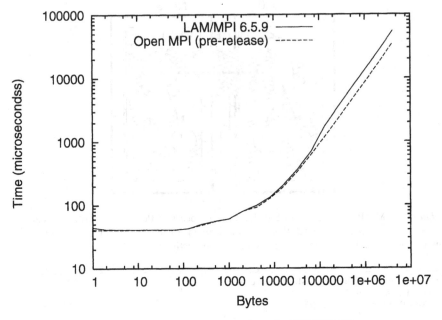

Figure 7.   Wall-clock execution times for MPI_BCAST.

timization technique is to switch between $O(n)$ and $O(log(n))$ when enough processes are involved in the collective. These collective algorithms were ported to the component architecture in Open MPI (the basic component, as described in Section 4.1). Measuring the performance of the same algorithms in two different architectures allows the comparison of overhead between the two.

A pair of dual-processor 2.0Ghz Intel Xeon nodes connected with Gigabit Ethernet and a dedicated switch was used for testing. Each node was running Red Hat 9 with Linux kernel 2.4.20 SMP and contained 2GB of RAM. The Pallas Benchmarks v2.2.1 were used to measure the wall-clock execution time of several MPI collectives in LAM/MPI and Open MPI.

The performance of MPI_BCAST, and MPI_ALLTOALL is shown in Figures 7 and 8, respectively. These graphs show that the performance of the collective algorithms in the Open MPI are on par with their peers in the LAM/MPI 6.5 series. Similarly, the performance of MPI_BARRIER is nearly identical between the two; wall-clock execution time for two processes was $73.5\mu s$ for LAM/MPI, and $80.2\mu s$ for Open MPI.

6.   Conclusions

Effective, easy-to-use tools for enabling research in high performance computing are critical to meet the ever-growing demands of scientific applications.

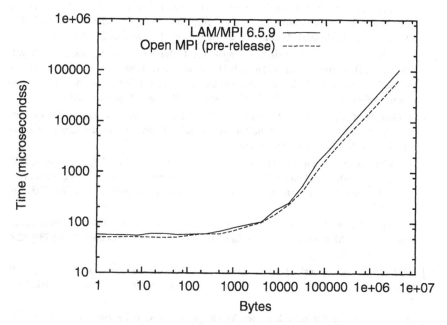

Figure 8. Wall-clock execution times for MPI_ALLTOALL.

The component framework of Open MPI allows third-party researchers to develop and test new algorithms within an MPI implementation without the large time investment required to first become an MPI implementor. This allows quicker development of algorithms as well as a robust vehicle to allow users access to cutting-edge research.

Future work includes completing and releasing Open MPI (expected November 2004), writing coll components to exploit high performance in a new environments, tighter integration of MPI topology-based communicators with collective algorithms, and continued development and integration of other component types within the Open MPI implementation (particularly as they relate to collective algorithms).

References

[1] R.T. Aulwes, D.J. Daniel, N.N. Desai, R.L. Graham, L.D. Risinger, M.W. Sukalski, M.A. Taylor, and T.S. Woodall. Architecture of LA-MPI, a network-fault-tolerant MPI. In *Proceedings of IPDPS'04*, April 2004.

[2] G. Burns, R. Daoud, and J. Vaigl. LAM: An Open Cluster Environment for MPI. In *Proceedings of Supercomputing Symposium*, pages 379–386, 1994.

[3] G.E. Fagg, A. Bukovsky, and J.J. Dongarra. HARNESS and fault tolerant MPI. *Parallel Computing*, 27:1479–1496, 2001.

[4]   G.E. Fagg, E. Gabriel, Z. Chen, T. Angskun, G. Bosilca, A. Bukovski, and J.J. Dongarra. Fault Tolerant Communication Library and Applications for High Perofrmance. In *Los Alamos Computer Science Institute Symposium*, Santa Fe, October 27-29 2003.

[5]   E. Gabriel, G.E. Fagg, G. Bosilca, T. Angskun, J.J. Dongarra, J.M. Squyres, V. Sahay, P. Kambadur, B. Barrett, A. Lumsdaine, R.H. Castain, D.J. Daniel, R.L. Graham, and T.S. Woodall. Open MPI: Goals, Concept, and Design of a Next Generation MPI Implementation. In *Proceedings, Euro PVM/MPI*, Budapest, Hungary, September 2004.

[6]   A. Geist, W. Gropp, S. Huss-Lederman, A. Lumsdaine, E. Lusk, W. Saphir, T. Skjellum, and M. Snir. MPI-2: Extending the Message-Passing Interface. In *Proceedings of Euro-Par'96*, LNCS, 1123:128–135, Springer, 1996.

[7]   R.L. Graham, S.E. Choi, D.J. Daniel, N.N. Desai, R.G. Minnich, C.E. Rasmussen, L.D. Risinger, and M.W. Sukalksi. A Network-failure-tolerant Message-passing System for Terascale Clusters. *International Journal of Parallel Programming*, 31(4):285–303, August 2003.

[8]   W. Gropp, E. Lusk, N. Doss, and A. Skjellum. A High-performance, Portable Implementation of the MPI Message Passing Interface Standard. *Parallel Computing*, 22(6):789–828, September 1996.

[9]   W.D. Gropp and E. Lusk. *User's Guide for* mpich, *a Portable Implementation of MPI*. Mathematics and Computer Science Division, Argonne National Laboratory, ANL-96/6, 1996.

[10] N. Karonis, B. de Supinski, I. Foster, W. Gropp, E. Lusk, and J. Bresnahan. Exploiting Hierarchy in Parallel Computer Networks to Optimize Collective Operation Performance. In *Proceedings of IPDPS'00*, pages 377–84, May 2000.

[11] A. Karwande, X. Yuan, and D. Lowenthal. CCMPI: A Compiled Communication Capable MPI Prototype for Ethernet Switched Clusters. In *ACM SIGPLAN Symposium on Principles and Practice of Parallel Programming*, San Diego, June 2003.

[12] R. Keller, E. Gabriel, B. Krammer, M.S. Müller, and M.M. Resch. Towards Efficient Execution of MPI Applications on the Grid: Porting and Optimization Issues. *Journal of Grid Computing*, 1(2):133–149, 2003.

[13] T. Kielmann, H.E. Bal, and S. Gorlatch. Bandwidth-efficient Collective Communication for Clustered Wide Area Systems. In *Proceedings of IPDPS'00*, pages 492–499, May 2000.

[14] T. Kielmann, R.F.H. Hofman, H.E. Bal, A. Plaat, and R.A.F. Bhoedjang. MagPIe: MPI's Collective Communication Operations for Clustered Wide Area Systems. *ACM SIGPLAN Symposium on Principles and Practice of Parallel Programming (PPoPP'99)*, pages 131–140, May 1999.

[15] S. P. Kini, J. Liu, J. Wu, P. Wyckoff, and D. K. Panda. Fast and Scalable Barrier Using RDMA and Multicast Mechanisms for InfiniBand-Based Cluster. In *Proceedings of Euro PVM/MPI*, LNCS, 2840, Springer, 2003.

[16] J.M. Mellor-Crummey and M.L. Scott. Algorithms for Scalable Synchronization on Shared-memory Multiprocessors. *ACM Transactions on Computer Systems*, 9(1):21–65, 1991.

[17] Message Passing Interface Forum. MPI: A Message Passing Interface. In *Proc. of Supercomputing '93*, pages 878–883. IEEE Computer Society Press, November 1993.

[18] S. Sankaran, J.M. Squyres, B. Barrett, A. Lumsdaine, J. Duell, P. Hargrove, and E. Roman. The LAM/MPI Checkpoint/Restart Framework: System-initiated Checkpointing. In *Proceedings of LACSI Symposium*, Sante Fe, October 2003.

[19] J.M. Squyres and A. Lumsdaine. A Component Architecture for LAM/MPI. In *Proceedings of Euro PVM/MPI*, LNCS, 2840, Springer, 2003.

[20] C. Szyperski, D. Druntz, and S. Murer. *Component Software: Beyond Object-Oriented Programming*. Addison Wesley, second edition, 2002.

[21] R. Thakur and W. Gropp. Improving the Performance of MPI Collective Communication on Switched Networks. Technical report ANL/MCS-P1007-1102, Mathematics and Computer Science Division, Argonne National Laboratory, November 2002. ftp://info.mcs.anl.gov/pub/tech_reports/reports/P1007.pdf.

[22] R. Thakur and W. Gropp. Improving the Performance of Collective Operations in MPICH. In *Proceedings of Euro PVM/MPI*, LNCS, 2840, Springer, 2003.

[23] T.S. Woodall, R.L. Graham, R.H. Castain, D.J. Daniel, M.W. Sukalski, G.E. Fagg, E. Gabriel, G. Bosilca, T. Angskun, J.J. Dongarra, J.M. Squyres, V. Sahay, P. Kambadur, B. Barrett, and A. Lumsdaine. TEG: A High-performance, Scalable, Multi-network Point-to-point Communications Methodology. In *Proceedings of Euro PVM/MPI*, Budapest, Hungary, September 2004.

[24] Q. Zhang. MPI Collective Operations Over Myrinet. Master's thesis, The University of British Columbia, Department of Computer Science, June 2002.

Index